Modern Physics

Modern Physics

Nathaniel Wren

NY RESEARCH
PRESS
New York

Published by NY Research Press
118-35 Queens Blvd., Suite 400,
Forest Hills, NY 11375, USA
www.nyresearchpress.com

Modern Physics
Nathaniel Wren

International Standard Book Number: 978-1-63238-702-8 (Hardback)

Cataloging-in-Publication Data

Modern physics / Nathaniel Wren.
p. cm.
Includes bibliographical references and index.
ISBN 978-1-63238-702-8
1. Physics. 2. Physical sciences. 3. Dynamics. I. Wren, Nathaniel.
QC6 .M63 2019
530.01--dc23

Printed in China.

Contents

Preface

The physics that integrates the principles of quantum mechanics and Einsteinian relativity is known as modern physics. It is crucial for understanding phenomena at extreme conditions. Quantum effects are exhibited at distances comparable to the size of atoms, and relativistic effects are seen for velocities approaching the speed of light. Although relativistic and quantum effects are present across all scales, the effects are often negligible. The concepts of black-body radiation, Franck-Hertz experiment, photoelectric effect, Stern-Gerlach experiment, etc. are fundamental to the development of modern physics from a framework of quantum physics. Similarly, relativistic physics develops from an understanding of the Michelson-Morley experiment, gravitational lensing and the perihelion precession of Mercury. This book provides comprehensive insights into the field of modern physics. It elucidates the concepts and innovative models around prospective developments with respect to this field. For someone with an interest and eye for detail, this book covers the most significant topics in modern physics.

To facilitate a deeper understanding of the contents of this book a short introduction of every chapter is written below:

Chapter 1- Modern physics is an approach to the understanding of nature that involves a quantum mechanical or relativistic description of nature. This chapter has been carefully written to provide an easy understanding of the varied concepts of modern physics, such as space, time, energy, entropy, particle, etc.

Chapter 2- Quantum mechanics is a fundamental theory of physics that is concerned with the study of phenomena at the smallest energy scales of atomic and subatomic particles. The aim of this chapter is to explore the key concepts of quantum mechanics including superposition, tunneling, quantum entanglement, wave function, wave particle duality, uncertainty principle, etc.

Chapter 3- An important area of physics is relativistic mechanics, which studies mechanics that is compatible with general and special relativity. This chapter has been carefully written to provide an easy understanding of the important concepts of relativistic mechanics, special relativity and general relativity, and includes vital topics such as time dilation, Lorentz transformation, relativistic space-time, relativistic mass, momentum and energy, etc.

Chapter 4- In physics, statistical mechanics is the approach to the study of a system with many degrees of freedom. This approach is based on an understanding of probability theory, statistical methods and physical laws. In order to completely understand statistical mechanics, it is necessary to study the concepts of microstate and macrostate, statistical ensemble, partition function, etc. which have been extensively discussed in this chapter.

Chapter 5- Nuclear physics studies the nucleus, its structure and interactions. The concepts of nuclear structure, stability of the nucleus, nuclear spin, radioactivity, etc. are essential for the development of the subject. Various other topics such as nuclear magnetic resonance, nuclear reactors, nuclear reactions, etc. have also been extensively covered for an extensive understanding of the subject.

I owe the completion of this book to the never-ending support of my family, who supported me throughout the project.

Nathaniel Wren

1

Fundamentals of Modern Physics

Modern physics is an approach to the understanding of nature that involves a quantum mechanical or relativistic description of nature. This chapter has been carefully written to provide an easy understanding of the varied concepts of modern physics, such as space, time, energy, entropy, particle, etc.

Modern physics is a branch of physics in which matter and energy are not separate, but instead are alternate forms of one another. The term modern physics refers to the study of facts and theories developed in this century that concern the interactions of matter, space and time. There have been enormous advances in physics, and some of these advances have brought into question, or have directly contradicted, certain theories that had seemed to be strongly supported by the experimental evidence.

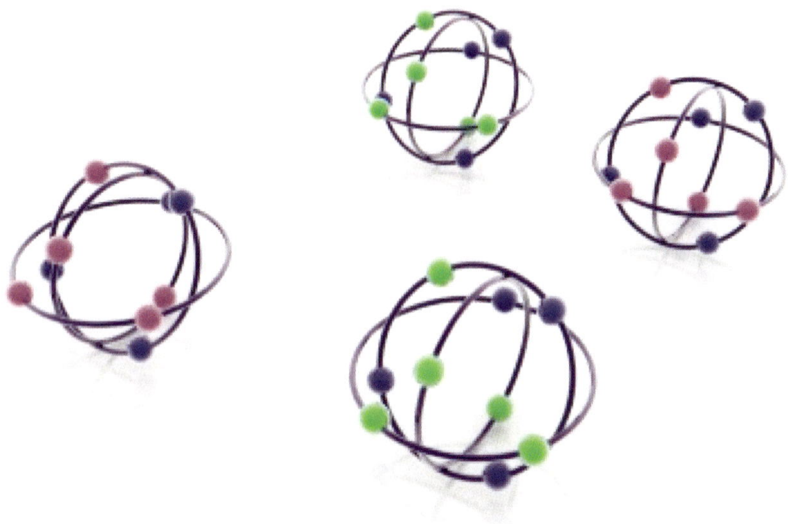

Modern physics is a rich field including decisive experiments conducted in the early part of the twentieth century and most recent research that has given us a deeper understanding of fundamental process in nature. In conjunction with our growing understanding of the physical world, a burgeoning technology has led to the development of lasers, solid state devices, and many other innovations.

Father of Modern Physics

Modern physics is a branch of physics existing after Newton. The classical physics was not enough to solve and explain the existing problems and hence modern physics came into existence. Einstein is called the father of modern physics. He was the one who came up with most of the concepts of modern physics. For example the theory of relativity was developed by him. Albert Einstein turned the scientific world on its head when he figured out E=mc², the formula that said that energy and mass were actually the same thing i.e. the ability to do work and the material to do work with.

Einstein's equation,

$$E = mc^2$$

Einstein changed the way we think about space, time and matter. His contribution to theoretical physics was significant enough to award him with Nobel Prize. So he is rightly justified as the Father of Modern Physics.

Particle Physics

Particle physics is the study of the fundamental constituents of matter and the forces of nature. This branch of science deals with the particles that make up matter. The interaction between these particles and their existence is studied in particle physics. The standard model is the latest model that describes the existence of particles that make up matter, any some fundamental interaction amongst them. These are the strong, weak as well as electromagnetic interactions. It states the existence of 24 fundamental particles and predicts the existence of Higgs Boson. The term "particle physics" usually refers to the study of "smallest" particles and the fields that must be defined in order to explain the observed particles which are the constituents of matter. Particle physics aims to develop fundamental models, conduct experiments etc. for the discovery of new fundamental particles that make up matter.

The main aim is to develop and research beyond the "standard model" so that further revolutions can be made. There include Collider and non-Collider experiments.

Nuclear Physics

Nuclear Physics is a branch of science that deals with the study of fundamental particles namely nucleus, the constituents and interactions of atomic nuclei. Nuclei are made up of protons and neutrons bound together by the strong force. Both protons and neutrons are referred to as nucleons. A heavy nucleus can contain hundreds of nucleons wherein the nucleus has an energy which arises partly from surface tension and partly from electrical repulsion of the protons. Nuclear physics deals with the study of structure and components of nucleus. Various models related to the nucleus and their development by various scientists, their experiments, theories etc. are part of Nuclear Physics.

Birth of Modern Physics

The State of Physics at the End of the 19th Century

By the end of the nineteenth century, already a large amount was known concerning the behaviour of objects subjected to forces of different kinds. But it had been a long and arduous road on the acquisition of this knowledge, with contributions from several great scientists. For example, the laws of mechanics began to really take shape with the work of Galileo Galilei (1564-1642), who performed several experiments that served as inspirations for generations of scientists that followed. He was the first, for example, to use the telescope to systematically study the skies and reveal the motions of planets and their satellites. These observations are closely related to the work of Johannes Kepler (1571-1630) who derived the so-called Kepler's Laws to describe the motions of planets within our solar system. It is not a coincidence nor a surprise that these empirical laws where later explained by Isaac Newton (1642-1727), as he is perhaps the greatest physicist of all time (not to neglect his fundamental contributions to several fields of mathematics).

But the state of knowledge at the end of the nineteenth century was not simply limited to the laws of mechanics or their application to astronomy, for example. There was tremendous progress that followed with the development of two important theories whose formulation culminated in the latter half of the 1800s: electromagnetism and thermodynamics. We now very briefly summarize these theories (including mechanics) as they were understood at that time.

Mechanics - Newton's Laws

It was Isaac Newton who first introduced the concepts of mass and force, to a large extent make sense of the experimental results obtained by previous scientists. Using these concepts, or principles, he was able to put forth three fundamental laws of motions (i.e., Newton's Laws of Motion) upon which much of classical physics rests:

i. A body remains at rest or in uniform motion unless acted upon by a net force.

ii. A body acted upon by a net force moves in such a manner that the time rate of change of the momentum equals the force.

iii. If two bodies exert forces on each other, these forces are equal in magnitude and opposite in direction.

The First Law would be meaningless without the concept of force, but it conveys a precise meaning for the concept of a "zero net force". This tendency for a body to remain in its initial state of motion (or at rest) is called inertia. One should note that according to the first law, the only thing that matters is the resultant force. It is irrelevant whether two or three forces (or any number for that matter) are applied simultaneously, if they cancel each other out, then their effect (or lack there of) is the same as that of having no force at all applied to the body. We then say that the body is in equilibrium.

The Second Law is very explicit: Force is the time rate of change of the momentum,

$$\mathrm{p} \equiv m\mathrm{v},$$

with m the mass, and v the velocity of the body. We therefore rewrite the Second Law as,

$$\begin{aligned} \mathrm{F}_{\mathrm{net}} &= \frac{dp}{dt} \\ &= \frac{d}{dt}(m\mathrm{v}), \end{aligned}$$

Although we still do not have a definition for the concept of mass, we can further transform above equation, if we assume that it is a constant, to yield,

$$F_{net} = m\frac{dv}{dt}$$
$$= ma,$$

with $a \equiv dv/dt$ the acceleration resulting from the action of the net force on the body.

The concept of mass is made clear with the Third Law. If we have two isolated bodies, 1 and 2, then the Third Law states that,

$$F_1 = -F_2,$$

and from the Second Law we have,

$$\frac{dp}{dt} = -\frac{dp_2}{dt},$$

or using the acceleration **a,**

$$m_1 a_1 = -m_2 a_2$$
$$\frac{m_1}{m_2} = \frac{a_2}{a_1},$$

with $a_i = |a_i|$. If one chooses m_1 as the reference or unit mass, m_2, or the mass of any other object, can be measured by comparison (if it is allowed to interact with m1) of their measured accelerations. Incidentally, we can use $\frac{d\mathbf{p}}{dt} = -\frac{d\mathbf{p}_2}{dt}$, to provide a different interpretation of Newton's Second Law,

$$\frac{d}{dt}(p_1 + p_2) = 0$$

Or

$$p_1 + p_2 = constant.$$

The momentum is conserved in the interaction of two isolated particles. This is a special case of the conservation of linear momentum. Also resulting from Newton's Laws are the conservation of angular momentum and the conservation of energy. These are important concepts that carry over to all other theories, classical or modern. As far as classical mechanics is concerned, the conservation of mass can also play a fundamental role in understanding the physical world.

Example – Kepler's Second Law from Newton's Laws and the conservation of angular momentum.

Newton's accomplishments are numerous and far-reaching. The aforementioned three laws are far from the only things he is known for. The derivation or explanation of Kepler's Laws of Planetary

Motion from the combination of Newton's Laws and his universal theory of gravitation is certainly a salient result of classical physics.

For Kepler's Second Law in particular, we can start by considering the torque acting on a planet orbiting the Sun,

$$\tau = \mathrm{r} \times \mathrm{F}_{\mathrm{grav},}$$

where $\tau = d\mathrm{L} / dt$ with L the angular momentum. However, since the vector r defining the position of the planet in relation to the Sun and the gravitational pull of the sun on the planet $\mathrm{F}_{\mathrm{grav}}$ are collinear, the torque on the planet is zero. It follows that the angular momentum of the planet is conserved.

Considering the definition for the angular momentum,

$$\begin{aligned} \mathrm{L} &= \mathrm{r} \times \mathrm{p} \\ &= mrv\sin(\phi)\mathrm{e}_{\perp} \\ &= \text{constant}, \end{aligned}$$

where $\mathrm{e}_{\perp}$ is the unit vector perpendicular to the orbital plane, m is the planet's mass, and is the angle between planet's linear momentum vector p and r. Since the orbital motion takes place in a plane, we can use,

$$\begin{aligned} v_{\perp} &\equiv v\sin(\phi) \\ &= \omega r, \end{aligned}$$

to transform equation mentioned before the above one, to,

$$\begin{aligned} \frac{L}{m} &= r^2\omega \\ &= r^2\frac{d\theta}{dt}, \end{aligned}$$

since $\omega = d\theta / dt$. However, from Figure 1, the triangular area dA swept as the planet travels an infinitesimal angle $d\theta$ is one-half the product of the radius r and the arc $rd\theta$,

$$dA = \frac{1}{2}r^2 d\theta.$$

Equation mentioned before the above one, then becomes,

$$\frac{L}{2m} = \frac{dA}{dt},$$

but since the angular momentum is conserved the area swept per unit time is found to agree with Kepler's Second Law, i.e.,

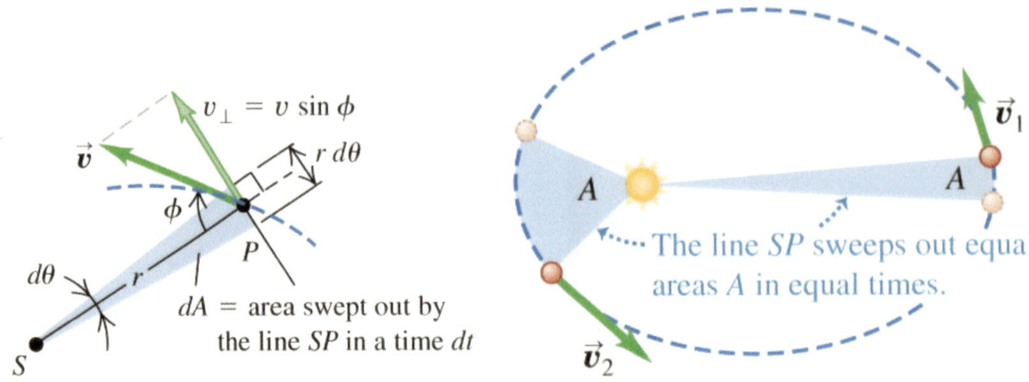

The area swept by a planet orbiting the sun in a given time interval.

$$\frac{dA}{dt} = \text{constant.}$$

It also follows that the total area A swept over a finite interval $T_2 - T_1$ is constant since,

$$\begin{aligned} A &= \int_{T_1}^{T_2} dA \\ &= \frac{L}{2m}\int_{T_1}^{T_2} dt \\ &= \frac{L}{2m}(T_2 - T_1). \end{aligned}$$

It therefore does not matter what part of the orbit the planet travels during the interval, the total area swept is the same.

Electromagnetism

Another spectacular success of classical physics is the discovery of the laws that regulate the behaviour of charged particles and currents, and their associated electric and magnetic fields. This was accomplished starting in the eighteenth century by several scientists (too numerous to name here) and culminated with the seminal work of James Clerk Maxwell (1831-1879). The laws that define electromagnetism can be succinctly summarized with the so-called Maxwell's Equations (using MKS units),

$$\begin{aligned} \nabla \cdot \mathrm{D} &= \rho \\ \nabla \cdot \mathrm{B} &= 0 \\ \nabla \times \mathrm{H} &= \mathrm{J} + \frac{\partial \mathrm{D}}{\partial t} \\ \nabla \times \mathrm{E} &= -\frac{\partial B}{\partial t}, \end{aligned}$$

where the electric displacement field D is related to the electric field E through $D = \varepsilon E$, with ε the electric permittivity, and the magnetic induction field B to the magnetic field $\mathrm{B} = \mu \mathrm{H}$, with μ the permeability. The quantity J in equations above, is the current density. These equations must be supplemented with the Lorentz Force (Hendrik Antoon Lorentz 1853-1928) that describes the dynamics of a particle of charge q interacting with the electromagnetic field,

$$\mathrm{F} = q(\mathrm{E} + \mathrm{v} \times \mathrm{B}).$$

These equations can be used to explain a whole set of phenomena, from the basic dynamical behaviour of, and the radiation from, an electron, for example, to the working of an electrical motor. These equations were also used by Maxwell to predict the propagation of electromagnetic waves and show that light is a special case of this phenomenon.

Example – The propagation of electromagnetic waves in free space.

We can study the behaviour of the electric field in free space, i.e., at a location away from any charge or current such that $\rho = J = 0$, and where $\varepsilon = \varepsilon_o$ and $\mu = \mu_o$ are the constant permittivity and permeability of vacuum. We proceed by taking the curl of the fourth of equations one before the above one,

$$\begin{aligned}\nabla\times(\nabla\times\mathrm{E})+\frac{\partial}{\partial t}(\nabla\times\mathrm{B}) &= \nabla(\nabla\cdot\mathrm{E})-\nabla^2\mathrm{E}+\frac{\partial}{\partial t}\left(\mu_0\varepsilon_0\frac{\partial\mathrm{E}}{\partial t}\right)\\ &= -\nabla^2\mathrm{E}+\frac{1}{c^2}\frac{\partial^2\mathrm{E}}{\partial t^2}\\ &= 0,\end{aligned}$$

where we have used the first and third of equations mentioned in the starting of the topic, the vector identity,

$$\nabla \times (\nabla \times \mathrm{E}) = \nabla(\nabla\cdot\mathrm{E}) - \nabla^2 \mathrm{E},$$

and defined a new constant,

$$c \equiv \frac{1}{\sqrt{\mu_0\varepsilon_0}}$$

The gradient operator is defined such that for a function *f* we have when using Cartesian coordinates,

$$\nabla f = \frac{\partial f}{\partial x} e_x + \frac{\partial f}{\partial y} e_y + \frac{\partial f}{\partial z} e_z,$$

where e_x , e_y and e_z are unit vectors in the corresponding directions. It also follows from this that for an arbitrary vector **a** the divergence, curl, and Laplacian are given by,

$$\nabla\cdot\mathrm{a} = \frac{\partial a_x}{\partial x} + \frac{\partial a_y}{\partial y} + \frac{\partial a_z}{\partial z}$$

$$\nabla \times a = \left(\frac{\partial a_z}{\partial y} - \frac{\partial a_y}{\partial z}\right) e_x + \left(\frac{\partial a_x}{\partial z} - \frac{\partial a_z}{\partial x}\right) e_y + \left(\frac{\partial a_y}{\partial x} - \frac{\partial a_x}{\partial y}\right) e_z$$

$$\nabla^2 f = \frac{\partial^2 f}{\partial x^2} + \frac{\partial^2 f}{\partial y^2} + \frac{\partial^2 f}{\partial z^2}.$$

If we simplify third equation from above, by considering the one-dimensional case where the electric field polarized in z is a function of, say, only the x direction, then we can write,

$$\frac{\partial^2 E_z}{\partial x^2} - \frac{1}{c^2}\frac{\partial^2 E_z}{\partial t^2} = 0,$$

for which it can be verified that the general solution is,

$$E_z = E_z\left(x \pm ct\right).$$

That is, the electric field is a function of the quantity x ± ct such that,

$$\frac{\partial^2 E_z}{\partial x^2} = AE_z$$

$$\frac{\partial^2 E_z}{\partial t^2} = c^2 AE_z,$$

with A some constant. Since the function,

$$X \equiv x \pm ct$$

can be interpreted as the position at time t for some physical quantity initially located at x when t = 0 , It therefore follows that equation $E_z = E_z\left(x \pm ct\right)$. corresponds to an electric field that travels either to the right (' – ' sign) or to the left ('+' sign) at the speed c. This can be made more apparent if we consider the particular solution,

$$E_z = E_0 \cos\left[k\left(x - ct\right)\right],$$

where we can readily see that a fixed phase,

$$\varphi = k\left(x - ct\right)$$

cannot and will not remain stationary at a location x for all times. For example, the case of $\varphi = 0$ will propagate (to the right) to successive positions $x = ct$.

Thermodynamics

Thermodynamics was established mainly as an empirical science that dealt with specific quantities pertaining to a given system. These quantities, which are used to specify the system's (physical or thermodynamic) state, can be mechanical in nature (e.g., pressure, volume, and work) while others are more subtle and needed the contributions of several scientists before a good understanding was attained (e.g., temperature, heat, and entropy).

The main results stemming from thermodynamics can be summarized by two fundamental laws:

The First Law of Thermodynamics: It generalizes the usual statement of conservation of mechanical energy encountered in Newtonian mechanics by also considering heat. It can be stated as follows: The change in internal energy U of a system equals to the heat Q added to it plus the work W done on the system. Mathematically this can written as,

$$\Delta U = Q + W .$$

The heat can thus be interpreted as the amount of energy received by the system in forms other than work.

The Second Law of Thermodynamics: Although the first law is essential in stating the conservation of energy (i.e., energy cannot be created out of nothing), it places no constraint on how energy can be transformed from one form into another. For example, although it is clear that an amount of work can be transformed into heat, can this heat be afterwards transform back into work? This question is addressed with the Second Law, which can be stated as follows (there are many different versions of this law): It is not possible to convert heat completely into work without some other change in the system taking place.

Example – The Specific Heat of an Ideal Gas.

Thermodynamics was very useful in establishing important results that could be tested experimentally. For example, the independent work from early scientists (Robert Boyle 1627-1691, Jacques Charles 1746-1823, Joseph Louis Gay-Lussac 1778-1850, and Amedeo Avogadro 1776-1856) could be combined to establish the equation of state for the a so-called ideal or perfect gas,

$$PV = nRT,$$

where P and V are the pressure and volume, n is the number of moles contained in the gas (one mole contains $N_A = 6.022 \times 10^{23}$ particles, with N_A the Avogadro Number, and $R = 8.31$ J/mol·K is the ideal gas constant. This relation can in turn be combined with equation $\Delta U = Q + W$. to calculate the heat capacity at constant pressure C_P or constant volume C_V, respectively. To do so, we first consider infinitesimal changes in our system such that equation $\Delta U = Q + W$. becomes,

$$dU = dQ + dW .$$

But the infinitesimal work done on the system can be written (since the pressure can be thought as energy per unit volume (as well as the force per unit area)),

$$dW = -PdV,$$

and thus,

$$dU + PdV = dQ.$$

For a given system containing n moles of gas, equation $PV \quad nRT$, makes it clear that the quantities P , V , and T cannot all be varied independently. For example, one could change the volume and the temperature, and the pressure would adjust itself to the new conditions. Alternatively, one could choose to vary the pressure and the temperature instead, while letting the volume expand or contract as needed, or vary the volume and the pressure with a resulting heating or cooling of the gas. In any case, one is left with a choice of two independent variables upon which to act. With T and V independent we can write the infinitesimal change in internal energy as,

$$dU = \left(\frac{\partial U}{\partial V}\right)_T dV + \left(\frac{\partial U}{\partial T}\right)_V dT,$$

and equation $dU + PdV = dQ.$ becomes,

$$\left(\frac{\partial U}{\partial T}\right)_V dT + \left[\left(\frac{\partial U}{\partial V}\right)_T + P\right] dV = dQ.$$

In equations above, the subscripts at the bottom of the parentheses indicate that the corresponding variable is to be held constant while the derivative is effected relative to the other independent variable.

Similarly, we can calculate a corresponding relation for dU when T and P are chosen as independent, but we must also ensure to express the other differential in equation $dU + PdV = dQ.$ as,

$$dV = \left(\frac{\partial V}{\partial T}\right)_P dT + \left(\frac{\partial V}{\partial P}\right)_T dP,$$

and therefore,

$$\left[\left(\frac{\partial U}{\partial T}\right)_P + P\left(\frac{\partial V}{\partial T}\right)_P\right] dT + \left[\left(\frac{\partial U}{\partial P}\right)_T + P\left(\frac{\partial V}{\partial P}\right)_T\right] dP = dQ.$$

Finally, taking V and P as independent variables we have,

$$\left(\frac{\partial U}{\partial P}\right)_V dP + \left[\left(\frac{\partial U}{\partial V}\right)_P + P\right] dV = dQ.$$

We can now use equations $\left(\frac{\partial U}{\partial T}\right)_V dT + \left[\left(\frac{\partial U}{\partial V}\right)_T + P\right] dV = dQ.$ and the one mentioned before above to find expressions for the heat capacities,

$$\begin{aligned} C_v &\equiv \left(\frac{\partial Q}{\partial T}\right)_V \\ &= \left(\frac{\partial U}{\partial T}\right)_v \\ C_P &\equiv \left(\frac{\partial Q}{\partial T}\right)_P \\ &= \left(\frac{\partial U}{\partial T}\right)_P + P\left(\frac{\partial V}{\partial T}\right)_P. \end{aligned}$$

For the case of an ideal gas it was shown experimentally by James Joule (1818-1889) that the internal energy of the gas is completely independent of the volume4 . It therefore follows that

$$C_V = \frac{dU}{dT},$$

and, from equation $dU + PdV = dQ.$,

$$C_V\, dT + PdV = dQ.$$

Then using equation $PV = nRT$, for one mole of an ideal gas (i.e., n = 1; the heat capacities are then called molar heat capacities) at constant pressure, the last of equations before the two mentioned just above becomes,

$$C_P = C_V + R.$$

These relations could be verified experimentally and were found to work well for approximately ideal gases. But with the acceptance of the Atomic Theory of Matter it became possible to derive such equations (and much of thermodynamics, for that matter) through Statistical Mechanics and the Kinetic Theory of Gases. In fact, discrepancies between predicted and experimental values for the molecular heats were one clue that all was not perfect with classical physics at the end of the nineteenth century.

Statistical Mechanics

The statistical interpretation of thermodynamics, through Statistical Mechanics (or Statistical Physics), was establish in the second half of the nineteenth century by Maxwell, Ludwig Eduard Boltzmann (1844-1906), and the American physicist J. Willard Gibbs (1839-1903). This theory bring us closer to the world of Modern Physics to come at the turn of the twentieth century in that it is based on the atomic theory of matter. Although atoms and molecules were still not well understood at the time (that would have to await the advent of Quantum Mechanics), the statistical interpretation of these constituents of matter would allow the calculation of important physical quantities and relations.

Example – The Maxwell-Boltzmann Distribution Law and the Equipartition of Energy.

In his study of statistical mechanics Maxwell considered the constituent of matter (i.e., atoms and molecules) as billiard balls interacting with each other and their environment through collisions. It was he, and with a later contribution from Boltzmann, who derived the velocity distribution for an ensemble of N identical particles (i.e., atoms or molecules) composing a gas for which the total energy is fixed. This is the Maxwell-Boltzmann Distribution Law,

$$f\left(v\right) = 4\pi N \left(\frac{m}{2\pi kT}\right)^{3/2} v^2 e^{-m\,v^2/\,2kT},$$

where m is the mass of a particle (the same mass for all), T the temperature of the gas, and $k \equiv R/N_A = 1.38 \times 10^{-23}\ J \cdot K^{-1}$ is the Boltzmann Constant. Alternatively, this distribution can be expressed using the speed $v = \left(v_2^x + v_y^2 + v_2^z\right)^{1/2}$ distribution with,

$$f\left(v\right) = 4\pi N\left(\frac{m}{2\pi kT}\right)^{3/2} v^2 e^{-mv2\ 2kT},$$

which is shown in figure below.The distributions are normalized such that,

$$N = \int_{-\infty}^{\infty}\int_{-\infty}^{\infty}\int_{-\infty}^{\infty} f(\mathbf{v})\,dv_x dv_y dv_z$$
$$= \int_0^{\infty} f(v)\,dv.$$

It is then possible to use these equations to calculate mean values for different quantities pertaining to the gas. For example, the mean kinetic energy of a particle due to its velocity in, say, the x -direction will be given by,

$$\overline{K}_x = \frac{1}{N}\int_{-\infty}^{\infty}\int_{-\infty}^{\infty}\int_{-\infty}^{\infty}\frac{1}{2}mv_x^2 f(\mathbf{v})dv_x dv_y dv_z$$
$$= \frac{m}{2}\left(\frac{m}{2\pi kT}\right)^{3/2}\left(\int_{-\infty}^{\infty} v_x^2 e^{-mv_x^2/2kT}dv_x\right)\left(\int_{-\infty}^{\infty} e^{-mv_y^2/2kT}dv_y\right)\left(\int_{-\infty}^{\infty} e^{-mv_z^2/2kT}dv_z\right)$$
$$= \frac{m}{2}\left(\frac{m}{2\pi kT}\right)^{3/2}\left(\int_{-\infty}^{\infty} v_x^2 e^{-mv_x^2/2kT}dv_x\right)\left(\int_{-\infty}^{\infty} e^{-mw^2/2kT}dw\right)^2$$
$$= \frac{m}{2}\left(\frac{m}{2\pi kT}\right)^{3/2}\left[\frac{1}{2\pi}\left(\frac{2\pi kT}{m}\right)^{3/2}\right]\left[\left(\frac{2\pi kT}{m}\right)^{1/2}\right]^2$$
$$= \frac{1}{2}kT$$

where the first integral on the second line was integrated by parts and we used above equation and equation mentioned before the above two ones, for the second. Performing a similar calculations for the y and z directions we find that the total mean kinetic energy for one particle is,

$$\overline{K} = \frac{3}{2}kT.$$

We can extend this result to a gas containing n moles for the total (internal) energy,

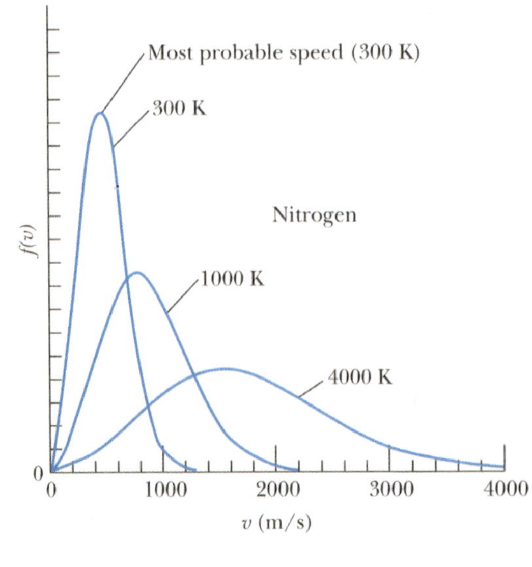

The Maxwell-Boltzmann distribution, i.e., equation $f(v) = 4\pi N\left(\frac{m}{2\pi kT}\right)^{3/2} v^2 e^{-mv2\ 2kT}$, for the nitrogen molecule at different temperatures

$$U = \frac{3}{2} nN_A kT,$$

since $N = nN_A$, and find from equation $C_V = \frac{dU}{dT}$, that the molar heat capacity (i.e., when n = 1) at constant volume,

$$C_V = \frac{3}{2} N_A k$$
$$= \frac{3}{2} R$$

We then find that Statistical Mechanics provides us with a prediction for a fundamental quantity that arose within the context of thermodynamics.

But it is important to note that the result obtained with equation mentioned before the equation above the graph for motion in one direction is a special case of the more general Equipartition Theorem of energy, which states that each degree of freedom of a particle has an average energy of $kT/2$ associated to it. The degrees of freedom of a system are defined as all the independent ways that energy can be stored in a system. For example, in the present case of a gas composed of colliding molecules considered here the only degrees of freedom arise from translational motions in three independent directions. That is, there are three such degrees.

On the other hand, a H_2 molecule made of two hydrogen atoms will have two more degrees of freedom associated to rotation about axes perpendicular to the axis joining the two nuclei, and another pair due to the possible vibration of the molecule from the stretching and compression of the bond joining the nuclei (for vibration there is one degree of freedom for the kinetic energy and another for the potential energy). We would then expect that a hydrogen molecule would have a heat capacity,

$$C_V(H_2) = (3 + 2 + 2) \cdot \frac{1}{2} N_A k$$
$$= \frac{7}{2} N_A k.$$

However, it is a characteristic of quantum mechanics that some degrees of freedom will only come into play when the energy of the system is high enough to excite them. For example, more energy is needed for the molecules in a hydrogen gas for rotational than translational motions, and even more for vibrations to take place. The resulting functional form of C_V/R as a function of the total energy of the gas (or the temperature) will then look like the curve shown in figure below. This "step-wise" behaviour of the molar heat capacity cannot be determined with classical mechanics, which would predict it to be a constant. It was Albert Einstein (1879-1955) who in 1907 first used the early ideas on energy quantization to improve on the results of classical physics and get accurate values.

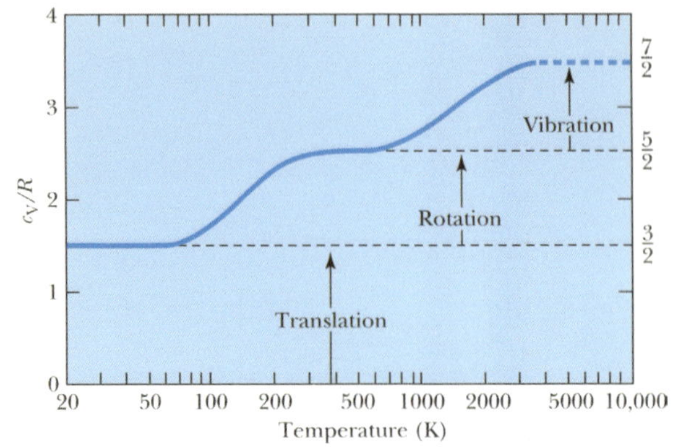

The molar heat capacity for a diatomic molecule, such as the molecular hydrogen, that exhibits translational, rotational, and vibrational degrees of freedom

for the heat capacity at high temperature. The Austrian physicist and chemist Peter Debye (1884-1966) eventually formulated a theory that correctly accounted for the observed behaviour at low temperature (as well as at high temperature) in 1912 (the socalled Debye Model).

Problems with Physics at the End of the 19th Century

We already saw in the previous section that thermodynamics and classical statistical mechanics (there is also a quantum mechanical version of statistical physics) could not adequately account for experimental measurements of heat capacities. This is one example of the limitations of classical mechanics, and it was eventually resolved by adopting the "new" ideas on the quantization of energy. But this was not the only problem facing classical physics at the end of the nineteen century, indeed there were a few much more serious crises.

The Ether, or the Electromagnetic Medium

Maxwell was able to show using the newly establish theory of electromagnetism that electromagnetic waves could propagate in space. Although not exactly the same, electromagnetic waves have much in common with acoustic or sound waves. In fact, our simple derivation of the wave equation seems to imply that electromagnetic waves should propagate in vacuum. This, of course, went completely against anything physicists of that period could imagine. After all, acoustic waves on a pond need the water to propagate, just like sound waves need the air in order to be heard. How could electromagnetic waves propagate on nothing?

To avoid this apparent inconsistency, physicists then postulated that an electromagnetic medium must exist on which waves can propagate. This medium, which was assumed to permeate all of space, was named the ether. But nature does not have to yield to human beliefs, and even the most sensitive experiments (most notably that of Albert Michelson (1852-1931) and Edward Morley (1838-1923) in 1887) could not detect the presence of the ether. The postulated electromagnetic medium did not exist.

The ether problem was solve by Einstein in 1905 when he published his Theory of Special Relativity, which is partly based on the postulate (and evidence from the Michelson-Morley experiment) that the speed of light is independent of the speed of the source from which it emanates. The notion of an electromagnetic medium was then shown to superfluous and not needed.

The Inertial Frames of Electromagnetism

Newtonian mechanics was in part build on the notion that there exit inertial frames of reference on which the laws of physics are the same. More precisely, an inertial frame is one that is in a uniform and non-accelerating state of motion. In fact, Newton's Laws are only applicable to such frame of references and his First Law can be used to define what an inertial frame is. It is often stated that Newton's Laws are invariant when moving from one inertial frame to another. This can be better understood mathematically.

If a particle of mass m has a velocity u′ relative to an observer who is at rest in an inertial frame K′, while this frame is moving with a constant velocity v as seen by another observer at rest in another inertial frame K (note the necessity of having a constant relative velocity v), then we would expect that the velocity u of the particle as measured in K to be,

$$u = u' + v.$$

That is, it would seem reasonable to expect that velocities should be added together when transforming from one inertial frame to another[5] ; such a transformation is called a Galilean transformation. If we write the mathematical form of the Second Law in frame K we have,

$$F = m\frac{du}{dt}$$

$$== m\,\frac{d(u' + v)}{dt},$$

but since v is constant,

$$F = m\frac{du'}{dt}F'$$

The result expressed through equation above is a statement of the invariance (or covariance) of Newton's Second Law. More precisely, the Second Law retains the same mathematical form no matter which inertial frame is used to express it, as long as velocities transform according to the simple addition rule stated in equation $u = u' + v$.

Another apparent problem with the theory of electromagnetism is that Maxwell's equations are not invariant to a Galilean transformation. That is, these equations change mathematical form when going from one Newtonian inertial frame to another, such as when using equation $u = u' + v$. It was natural for physicists of the nineteenth century to assume that Newton was right and Maxwell wrong. But electromagnetism seemed to be a very successful theory that agreed with experiments.

This problem was also eventually solved by Einstein's Theory of Special Relativity of 1905, where he showed that it was the other way around: Maxwell was right and Newton wrong. As it turns, Galilean transformations do not define inertial frames. Instead, the socalled Lorentz Transformation, previously and independently derived by Lorentz and the Irish physicist George FitzGerald (1851-1901), is the fundamental transformation that plays such a role (this transformation was also re-derived from scratch by Einstein in 1905). It followed that electromagnetism was found to be a theory perfectly consistent with special relativity. Newtonian mechanics, on the other hand, has shortcomings that are overcome by special relativity.

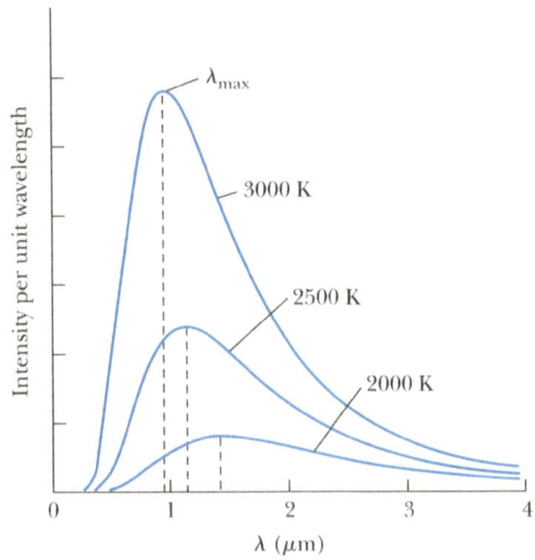

The blackbody spectrum as a function of wavelength for three different temperature

Blackbody Radiation

A blackbody is an object that absorbs all incident radiation and emits a continuous spectrum of radiation whose distribution and intensity depend only on its temperature (e.g., it does not depend on the nature of the body itself). Classical physics was unable to explain this spectrum. Although it was possible to explain the blackbody spectrum at low and high frequencies, no single theory could account for the shape of the whole spectrum.

This all changed in 1900 when Max Planck (1858-1947) introduced the notion of quantization for the first time. Not only was he able to solve the blackbody problem, his idea was revolutionary and served as the foundation for the development of Quantum Mechanics in the decades to follow.

Space

From the perspective of an Earthling, outer space is a zone that occurs about 100 kilometers (60 miles) above the planet, where there is no appreciable air to breathe or to scatter light. In that area, blue gives way to black because oxygen molecules are not in enough abundance to make the sky blue.

Further, space is a vacuum, meaning that sound cannot carry because molecules are not close enough together to transmit sound between them. That's not to say that space is empty, however. Gas, dust and other bits of matter float around "emptier" areas of the universe, while more crowded regions can host planets, stars and galaxies.

No one knows exactly how big space is. The difficulty arises because of what we can see in our detectors. We measure long distances in space in "light-years," representing the distance it takes for light to travel in a year (roughly 5.8 trillion miles, or 9.3 trillion kilometers).

From light that is visible in our telescopes, we have charted galaxies reaching almost as far back as the Big Bang, which is thought to have started our universe 13.7 billion years ago. This means we can "see" into space at a distance of almost 13.7 billion light-years. However, astronomers are not sure if our universe is the only universe that exists. This means that space could be a lot bigger than it appears to us.

Radiation Invisible to Human Eyes

Most of space is relatively empty, meaning that there are just stray bits of dust and gas inside of it. This means that when humans send a satellite to a distant planet, the object will not encounter "drag" in the same way that an airplane does as it sails through space.

The vacuum environment in space and on the moon, for example, is one reason that the lunar lander of the Apollo program looks so odd-shaped — like a spider, one crew said. Because the spacecraft was designed to work in a zone with no atmosphere, there was no need for smooth edges or an aerodynamic shape.

While space may look empty to human eyes, research has shown that there are forms of radiation emanating through the cosmos. In our own solar system, the solar wind — made up of plasma and other particles from the sun — permeates past the planets and occasionally causes aurora near the Earth's poles. Cosmic rays also fly through the neighborhood, emanating from supernovas outside of the solar system.

In fact, the universe is permeated with the cosmic microwave background, which can be understood as the leftovers of the immense explosion that formed our cosmos (usually called the Big Bang). The CMB, which is best seen in microwaves, shows the earliest radiation that our instruments can detect.

One large feature of space that is poorly seen or understood is the supposed presence of dark matter and dark energy, which are essentially forms of matter and energy that can only be detected through their effects on other objects. Since the universe is expanding and accelerating in that expansion, that is seen as one key piece of evidence for dark matter. Another is gravitational lensing that occurs when light "bends" around a star from a distant background object.

In astronomy and cosmology, space is the vast 3-dimensional region that begins where the earth's atmosphere ends. Space is usually thought to begin at the lowest altitude at which satellites can maintain orbits for a reasonable time without falling into the atmosphere. This is approximately 160 kilometers (100 miles) above the surface. Astronomers may speak of interplanetary space (the space between planets in our solar system), interstellar space (the space between stars in our galaxy), or intergalactic space (the space between galaxies in the universe). Some scientists believe that space extends infinitely far in all directions, while others believe that space is finite but unbounded, just as the 2-space surface of the earth has finite area yet no beginning nor end.

Spatial Measurement

The measurement of *physical space* has long been important. Although earlier societies had developed measuring systems, the International System of Units, (SI), is now the most common system of units used in the measuring of space, and is almost universally used.

Currently, the standard space interval, called a standard meter or simply meter, is defined as the distance traveled by light in a vacuum during a time interval of exactly 1/299,792,458 of a second. This definition coupled with present definition of the second is based on the special theory of relativity in which the speed of light plays the role of a fundamental constant of nature.

Time

Physics is the only science that explicitly studies time, but even physicists agree that time is one of the most difficult properties of our universe to understand. Even in the most modern and complex physical models, though, time is usually considered to be an ontologically "basic" or primary concept, and not made up of, or dependent on, anything else.

In the sciences generally, time is simply what a clock reads, but this hides a whole host of different conceptions of time used in physics

In the sciences generally, time is usually defined by its measurement: it is simply what a clock reads. Physics in particular often requires extreme levels of precision in time measurement, which has led to the requirement that time be considered an infinitely divisible linear continuum, and not quantized (i.e. composed of discrete and indivisible units). With modern atomic time standards like TAI and UTC and ultra-precise atomic clocks, time can now be measured accurate to about 10^{-15} seconds, which corresponds to about 1 second error in approximately 30 million years.

But several different conceptions and applications of time have been explored over the centuries in different areas of physics.

In non-relativistic or classical physics, the concept of time generally used is that of absolute time (also called Newtonian time after its most famous proponent), time which is independent of any perceiver, progresses at a consistent pace for everyone everywhere throughout the universe, and is essentially imperceptible and mathematical in nature. This accords with most people's everyday experience of how time flows.

However, since the advent of relativity in the early 20th Century, relativistic time has become the norm within physics. This takes into account phenomena such as time dilation for fast-moving objects, gravitational time dilation for objects caught in extreme gravitational fields, and the important idea that time is really just one element of four-dimensional space-time.

Relativity also allows for, at least in theory, the prospect of time travel, and there are several scenarios which allow for the theoretical basis of travel in time. There are even theoretical faster-than-light time-travelling particles like tachyons and neutrinos. However, the concept of time travel also brings with it a number of paradoxes, and its likelihood and physical practicality is questioned by many physicists.

Quantum mechanics revolutionized physics in the first half of the 20th Century and it still represents the most complete and accurate model of the universe we have. Time is perhaps not as central a concept in quantum theory as it is in classical physics, and there is really no such thing as "quantum time" as such. For example, time does not appear to be divided up into discrete quanta as are most other aspects of reality. However, the different interpretations of quantum theory (e.g. the Copenhagen interpretation, the many worlds interpretation, etc) do have some potentially important implications for our understanding of time.

Most physicists agree that time had a beginning, and that it is measured from, and indeed came into being with, The Big Bang some 13.8 billion years ago. Whether, how and when time might end in the future is a more open question, depending on different notions of the ultimate fate of the universe and other mind-bending concepts like the multiverse.

The so-called arrow of time refers to the one-way direction or asymmetry of time, which leads to the way we instinctively perceive time as moving forwards from the fixed and immutable past, though the present, towards the unknown and unfixed future. This idea has it roots in physics, particularly in the Second Law of Thermodynamics, although other, often related, arrows of time have also been identified.

Energy

Energy, in physics, the capacity for doing work. It may exist in potential, kinetic, thermal, electrical, chemical, nuclear, or other various forms. There are, moreover, heat and work—i.e., energy in the process of transfer from one body to another. After it has been transferred, energy is always designated according to its nature. Hence, heat transferred may become thermal energy, while work done may manifest itself in the form of mechanical energy.

All forms of energy are associated with motion. For example, any given body has kinetic energy if it is in motion. A tensioned device such as a bow or spring, though at rest, has the potential for creating motion; it contains potential energy because of its configuration. Similarly, nuclear energy is potential energy because it results from the configuration of subatomic particles in the nucleus of an atom.

Energy can be neither created nor destroyed but only changed from one form to another. This principle is known as the conservation of energy or the first law of thermodynamics. For example, when a box slides down a hill, the potential energy that the box has from being located high up on the slope is converted to kinetic energy, energy of motion. As the box slows to a stop through friction, the kinetic energy from the box's motion is converted to thermal energy that heats the box and the slope.

Energy can be converted from one form to another in various other ways. Usable mechanical or electrical energy is, for instance, produced by many kinds of devices, including fuel-burning heat engines, generators, batteries, fuel cells, and magnetohydrodynamic systems.

In the International System of Units (SI), energy is measured in joules. One joule is equal to the work done by a one-newton force acting over a one-metre distance.

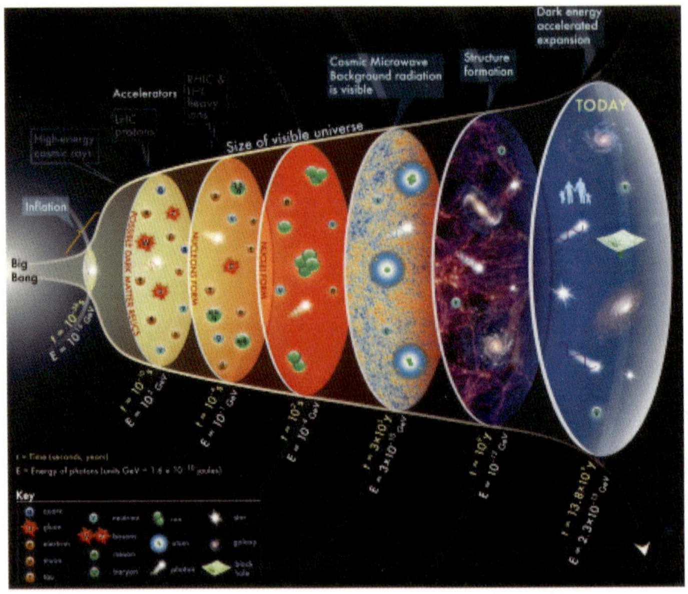

Classical Mechanics

In classical mechanics, energy is a conceptually and mathematically useful property, as it is a conserved quantity. Several formulations of mechanics have been developed using energy as a core concept.

Work, a function of energy, is force times distance.

$$W = \int_C \mathrm{F} \cdot \mathrm{ds}$$

This says that the work (W) is equal to the line integral of the force F along a path C. Work and thus energy is frame dependent. For example, consider a ball being hit by a bat. In the center-of-mass reference frame, the bat does no work on the ball. But, in the reference frame of the person swinging the bat, considerable work is done on the ball.

The total energy of a system is sometimes called the Hamiltonian, after William Rowan Hamilton. The classical equations of motion can be written in terms of the Hamiltonian, even for highly complex or abstract systems. These classical equations have remarkably direct analogs in nonrelativistic quantum mechanics.

Another energy-related concept is called the Lagrangian, after Joseph-Louis Lagrange. This formalism is as fundamental as the Hamiltonian, and both can be used to derive the equations of motion or be derived from them. It was invented in the context of classical mechanics, but is generally useful in modern physics. The Lagrangian is defined as the kinetic energy *minus* the potential energy. Usually, the Lagrange formalism is mathematically more convenient than the Hamiltonian for non-conservative systems (such as systems with friction).

Noether's theorem (1918) states that any differentiable symmetry of the action of a physical system has a corresponding conservation law. Noether's theorem has become a fundamental tool of modern theoretical physics and the calculus of variations. A generalisation of the seminal formulations on constants of motion in Lagrangian and Hamiltonian mechanics (1788 and 1833, respectively), it does not apply to systems that cannot be modeled with a Lagrangian; for example, dissipative systems with continuous symmetries need not have a corresponding conservation law.

Cosmology

In cosmology and astronomy the phenomena of stars, nova, supernova, quasars and gamma-ray bursts are the universe's highest-output energy transformations of matter. All stellarphenomena (including solar activity) are driven by various kinds of energy transformations. Energy in such transformations is either from gravitational collapse of matter (usually molecular hydrogen) into various classes of astronomical objects (stars, black holes, etc.), or from nuclear fusion (of lighter elements, primarily hydrogen). The nuclear fusion of hydrogen in the Sun also releases another store of potential energy which was created at the time of the Big Bang. At that time, according to theory, space expanded and the universe cooled too rapidly for hydrogen to completely fuse into heavier elements. This meant that hydrogen represents a store of potential energy that can be released by fusion. Such a fusion process is triggered by heat and pressure generated from gravitational collapse of hydrogen clouds when they produce stars, and some of the fusion energy is then transformed into sunlight.

Quantum Mechanics

In quantum mechanics, energy is defined in terms of the energy operator as a time derivative of the wave function. The Schrödinger equation equates the energy operator to the full energy of a particle or a system. Its results can be considered as a definition of measurement of energy in quantum mechanics. The Schrödinger equation describes the space- and time-dependence of a slowly changing (non-relativistic) wave function of quantum systems. The solution of this equation for a bound system is discrete (a set of permitted states, each characterized by an energy level) which results in the concept of quanta. In the solution of the Schrödinger equation for any oscillator (vibrator) and for electromagnetic waves in a vacuum, the resulting energy states are related to the frequency by Planck's relation: $E = h\nu$ (where h is Planck's constant and ν the frequency). In the case of an electromagnetic wave these energy states are called quanta of light or photons.

Relativity

When calculating kinetic energy (work to accelerate a massive body from zero speed to some finite speed) relativistically – using Lorentz transformations instead of Newtonian mechanics – Einstein discovered an unexpected by-product of these calculations to be an energy term which does not vanish at zero speed. He called it rest energy: energy which every massive body must possess even when being at rest. The amount of energy is directly proportional to the mass of the body:

$$E_0 = mc^2,$$

where,

m is the mass of the body,

c is the speed of light in vacuum,

E_0 is the rest energy.

For example, consider electron–positron annihilation, in which the rest energy of these two individual particles (equivalent to their rest mass) is converted to the radiant energy of the photons produced in the process. In this system the matter and antimatter (electrons and positrons) are destroyed and changed to non-matter (the photons). However, the total mass and total energy do not change during this interaction. The photons each have no rest mass but nonetheless have radiant energy which exhibits the same inertia as did the two original particles. This is a reversible process – the inverse process is called pair creation – in which the rest mass of particles is created from the radiant energy of two (or more) annihilating photons.

In general relativity, the stress–energy tensor serves as the source term for the gravitational field, in rough analogy to the way mass serves as the source term in the non-relativistic Newtonian approximation.

Energy and mass are manifestations of one and the same underlying physical property of a system. This property is responsible for the inertia and strength of gravitational interaction of the system ("mass manifestations"), and is also responsible for the potential ability of the system to perform work or heating ("energy manifestations"), subject to the limitations of other physical laws.

In classical physics, energy is a scalar quantity, the canonical conjugate to time. In special relativity energy is also a scalar (although not a Lorentz scalar but a time component of the energy–momentum 4-vector). In other words, energy is invariant with respect to rotations of space, but not invariant with respect to rotations of space-time (= boosts).

Transformation

Energy may be transformed between different forms at various efficiencies. Items that transform between these forms are called transducers. Examples of transducers include a battery, from chemical energy to electric energy; a dam: gravitational potential energy to kinetic energy of moving water (and the blades of a turbine) and ultimately to electric energy through an electric generator; or a heat engine, from heat to work.

Some forms of transfer of energy ("energy in transit") from one object or system to another

Type of transfer process	Description
Heat	that amount of thermal energy in transit spontaneously towards a lower-temperature object
Work	that amount of energy in transit due to a displacement in the direction of an applied force
Transfer of material	that amount of energy carried by matter that is moving from one system to another

A turbo generator transforms the energy of pressurised steam into electrical energy

Examples of energy transformation include generating electric energy from heat energy via a steam turbine, or lifting an object against gravity using electrical energy driving a crane motor. Lifting against gravity performs mechanical work on the object and stores gravitational potential energy in the object. If the object falls to the ground, gravity does mechanical work on the object which transforms the potential energy in the gravitational field to the kinetic energy released as heat on impact with the ground. Our Sun transforms nuclear potential energy to other forms of energy; its total mass does not decrease due to that in itself (since it still contains the same total energy even if in different forms), but its mass does decrease when the energy escapes out to its surroundings, largely as radiant energy.

There are strict limits to how efficiently heat can be converted into work in a cyclic process, e.g. in a heat engine, as described by Carnot's theorem and the second law of thermodynamics. However, some energy transformations can be quite efficient. The direction of transformations in energy (what kind of energy is transformed to what other kind) is often determined by entropy (equal energy spread among all available degrees of freedom) considerations. In practice all energy transformations are permitted on a small scale, but certain larger transformations are not permitted because it is statistically unlikely that energy or matter will randomly move into more concentrated forms or smaller spaces.

Energy transformations in the universe over time are characterized by various kinds of potential energy that has been available since the Big Bang later being "released" (transformed to more active types of energy such as kinetic or radiant energy) when a triggering mechanism is available. Familiar examples of such processes include nuclear decay, in which energy is released that was originally "stored" in heavy isotopes (such as uranium and thorium), by nucleosynthesis, a process ultimately using the gravitational potential energy released from the gravitational collapse of supernovae, to store energy in the creation of these heavy elements before they were incorporated into the solar system and the Earth. This energy is triggered and released in nuclear fission bombs or in civil nuclear power generation. Similarly, in the case of a chemical explosion, chemical potential energy is transformed to kinetic energy and thermal energy in a very short time. Yet another example is that of a pendulum. At its highest points the kinetic energy is zero and the gravitational potential energy is at maximum. At its lowest point the kinetic energy is at maximum and is equal to the decrease of potential energy. If one (unrealistically) assumes that there is no friction or other losses, the conversion of energy between these processes would be perfect, and the pendulum would continue swinging forever.

Energy is also transferred from potential energy (E_p) to kinetic energy (E_k) and then back to potential energy constantly. This is referred to as conservation of energy. In this closed system, energy cannot be created or destroyed; therefore, the initial energy and the final energy will be equal to each other. This can be demonstrated by the following:

$$E_{pi} + E_{ki} = E_{pF} + E_{kF}$$

The equation can then be simplified further since $E_p = mgh$ (mass times acceleration due to gravity times the height) and $E_k = \frac{1}{2}mv^2$ (half mass times velocity squared). Then the total amount of energy can be found by adding $E_p + E_k = E_{total}$.

Conservation of Energy and Mass in Transformation

Energy gives rise to weight when it is trapped in a system with zero momentum, where it can be weighed. It is also equivalent to mass, and this mass is always associated with it. Mass is also equivalent to a certain amount of energy, and likewise always appears associated with it, as described in mass-energy equivalence. The formula $E = mc^2$, derived by Albert Einstein (1905) quantifies the relationship between rest-mass and rest-energy within the concept of special relativity. In different theoretical frameworks, similar formulas were derived by J. J. Thomson (1881), Henri Poincaré (1900), Friedrich Hasenöhrl (1904) and others.

Part of the rest energy (equivalent to rest mass) of matter may be converted to other forms of energy (still exhibiting mass), but neither energy nor mass can be destroyed; rather, both remain constant during any process. However, since c^2 is extremely large relative to ordinary human scales, the conversion of an everyday amount of rest mass (for example, 1 kg) from rest energy to other forms of energy (such as kinetic energy, thermal energy, or the radiant energy carried by light and other radiation) can liberate tremendous amounts of energy ($\sim 9\times10^{16}$ joules = 21 megatons of TNT), as can be seen in nuclear reactors and nuclear weapons. Conversely, the mass equivalent of an everyday amount energy is minuscule, which is why a loss of energy (loss of mass) from most systems is difficult to measure on a weighing scale, unless the energy loss is very large. Examples of large transformations between rest energy (of matter) and other forms of energy (e.g., kinetic energy into particles with rest mass) are found in nuclear physics and particle physics.

Reversible and Non-reversible Transformations

Thermodynamics divides energy transformation into two kinds: reversible processes and irreversible processes. An irreversible process is one in which energy is dissipated (spread) into empty energy states available in a volume, from which it cannot be recovered into more concentrated forms (fewer quantum states), without degradation of even more energy. A reversible process is one in which this sort of dissipation does not happen. For example, conversion of energy from one type of potential field to another, is reversible, as in the pendulum system described above. In processes where heat is generated, quantum states of lower energy, present as possible excitations in fields between atoms, act as a reservoir for part of the energy, from which it cannot be recovered, in order to be converted with 100% efficiency into other forms of energy. In this case, the energy must partly stay as heat, and cannot be completely recovered as usable energy, except at the price of an increase in some other kind of heat-like increase in disorder in quantum states, in the universe (such as an expansion of matter, or a randomisation in a crystal).

As the universe evolves in time, more and more of its energy becomes trapped in irreversible states (i.e., as heat or other kinds of increases in disorder). This has been referred to as the inevitable thermodynamic heat death of the universe. In this heat death the energy of the universe does not change, but the fraction of energy which is available to do work through a heat engine, or be transformed to other usable forms of energy (through the use of generators attached to heat engines), grows less and less.

Conservation of Energy

The fact that energy can be neither created nor be destroyed is called the law of conservation of energy. In the form of the first law of thermodynamics, this states that a closed system's energy is constant unless energy is transferred in or out by work or heat, and that no energy is lost in transfer. The total inflow of energy into a system must equal the total outflow of energy from the system, plus the change in the energy contained within the system. Whenever one measures (or calculates) the total energy of a system of particles whose interactions do not depend explicitly on time, it is found that the total energy of the system always remains constant.

While heat can always be fully converted into work in a reversible isothermal expansion of an ideal gas, for cyclic processes of practical interest in heat engines the second law of thermodynamics states that the system doing work always loses some energy as waste heat. This creates a limit to the amount of heat energy that can do work in a cyclic process, a limit called the available energy. Mechanical and other forms of energy can be transformed in the other direction into thermal energy without such limitations. The total energy of a system can be calculated by adding up all forms of energy in the system.

Richard Feynman said during a 1961 lecture:

> "There is a fact, or if you wish, a *law*, governing all natural phenomena that are known to date. There is no known exception to this law—it is exact so far as we know. The law is called the *conservation of energy*. It states that there is a certain quantity, which we call energy, that does not change in manifold changes which nature undergoes. That is a most abstract idea, because it is a mathematical principle; it says that there is a numerical quantity which does not change when something happens. It is not a description of a mechanism, or anything concrete; it is just a strange fact that we can calculate some number and when we finish watching nature go through her tricks and calculate the number again, it is the same."
>
> — *The Feynman Lectures on Physics*

Most kinds of energy (with gravitational energy being a notable exception) are subject to strict local conservation laws as well. In this case, energy can only be exchanged between adjacent regions of space, and all observers agree as to the volumetric density of energy in any given space. There is also a global law of conservation of energy, stating that the total energy of the universe cannot change; this is a corollary of the local law, but not vice versa.

This law is a fundamental principle of physics. As shown rigorously by Noether's theorem, the conservation of energy is a mathematical consequence of translational symmetry of time, a property of most phenomena below the cosmic scale that makes them independent of their locations on the time coordinate. Put differently, yesterday, today, and tomorrow are physically indistinguishable. This is because energy is the quantity which is canonical conjugate to time. This mathematical entanglement of energy and time also results in the uncertainty principle - it is impossible to define the exact amount of energy during any definite time interval. The uncertainty principle should not be confused with energy conservation - rather it provides mathematical limits to which energy can in principle be defined and measured.

Each of the basic forces of nature is associated with a different type of potential energy, and all types of potential energy (like all other types of energy) appears as system mass, whenever present. For example, a compressed spring will be slightly more massive than before it was compressed. Likewise, whenever energy is transferred between systems by any mechanism, an associated mass is transferred with it.

In quantum mechanics energy is expressed using the Hamiltonian operator. On any time scales, the uncertainty in the energy is by,

$$\Delta E \Delta t \geq \frac{\hbar}{2}$$

which is similar in form to the Heisenberg Uncertainty Principle (but not really mathematically equivalent thereto, since H and t are not dynamically conjugate variables, neither in classical nor in quantum mechanics).

In particle physics, this inequality permits a qualitative understanding of virtual particles which carry momentum, exchange by which and with real particles, is responsible for the creation of all known fundamental forces (more accurately known as fundamental interactions). Virtual photons (which are simply lowest quantum mechanical energy state of photons) are also responsible for electrostatic interaction between electric charges (which results in Coulomb law), for spontaneous radiative decay of exited atomic and nuclear states, for the Casimir force, for van der Waals bond forces and some other observable phenomena.

Energy Transfer

Closed Systems

Energy transfer can be considered for the special case of systems which are closed to transfers of matter. The portion of the energy which is transferred by conservative forces over a distance is measured as the work the source system does on the receiving system. The portion of the energy which does not do work during the transfer is called heat. Energy can be transferred between systems in a variety of ways. Examples include the transmission of electromagnetic energy via photons, physical collisions which transfer kinetic energy, and the conductive transfer of thermal energy.

Energy is strictly conserved and is also locally conserved wherever it can be defined. In thermodynamics, for closed systems, the process of energy transfer is described by the first law:

$$\Delta E = W + Q$$

where E is the amount of energy transferred, W represents the work done on the system, and Q represents the heat flow into the system. As a simplification, the heat term, Q, is sometimes ignored, especially when the thermal efficiency of the transfer is high.

$$\Delta E = W$$

This simplified equation is the one used to define the joule, for example.

Open Systems

Beyond the constraints of closed systems, open systems can gain or lose energy in association with matter transfer (both of these process are illustrated by fueling an auto, a system which gains in energy thereby, without addition of either work or heat). Denoting this energy by E, one may write,

$$\Delta E = W + Q + E.$$

Thermodynamics

Internal Energy

Internal energy is the sum of all microscopic forms of energy of a system. It is the energy needed to create the system. It is related to the potential energy, e.g., molecular structure, crystal structure, and other geometric aspects, as well as the motion of the particles, in form of kinetic energy. Thermodynamics is chiefly concerned with changes in internal energy and not its absolute value, which is impossible to determine with thermodynamics alone.

First Law of Thermodynamics

The first law of thermodynamics asserts that energy (but not necessarily thermodynamic free energy) is always conserved and that heat flow is a form of energy transfer. For homogeneous systems, with a well-defined temperature and pressure, a commonly used corollary of the first law is that, for a system subject only to pressure forces and heat transfer (e.g., a cylinder-full of gas) without chemical changes, the differential change in the internal energy of the system (with a *gain* in energy signified by a positive quantity) is given as

$$\mathrm{d}E = T\mathrm{d}S - P\mathrm{d}V,$$

where the first term on the right is the heat transferred into the system, expressed in terms of temperature T and entropy S (in which entropy increases and the change $\mathrm{d}S$ is positive when the system is heated), and the last term on the right hand side is identified as work done on the system, where pressure is P and volume V (the negative sign results since compression of the system requires work to be done on it and so the volume change, $\mathrm{d}V$, is negative when work is done on the system).

This equation is highly specific, ignoring all chemical, electrical, nuclear, and gravitational forces, effects such as advection of any form of energy other than heat and pV-work. The general formulation of the first law (i.e., conservation of energy) is valid even in situations in which the system is not homogeneous. For these cases the change in internal energy of a *closed* system is expressed in a general form by,

$$\mathrm{d}E = \delta Q + \delta W$$

where δQ is the heat supplied to the system and δW is the work applied to the system.

Equipartition of Energy

The energy of a mechanical harmonic oscillator (a mass on a spring) is alternatively kinetic and potential. At two points in the oscillation cycle it is entirely kinetic, and at two points it is entirely potential. Over the whole cycle, or over many cycles, net energy is thus equally split between kinetic and potential. This is called equipartition principle; total energy of a system with many degrees of freedom is equally split among all available degrees of freedom.

This principle is vitally important to understanding the behaviour of a quantity closely related to energy, called entropy. Entropy is a measure of evenness of a distribution of energy between parts of a system. When an isolated system is given more degrees of freedom (i.e., given new available energy states that are the same as existing states), then total energy spreads over all available degrees equally without distinction between "new" and "old" degrees. This mathematical result is called the second law of thermodynamics. The second law of thermodynamics is valid only for systems which are near or in equilibrium state. For non-equilibrium systems, the laws governing system's behavior are still debatable. One of the guiding principles for these systems is the principle of maximum entropy production. It states that nonequilibrium systems behave in such a way to maximize its entropy production.

Entropy

The term entropy was first used by Rudolf Clausius to state the second law of thermodynamics.

Entropy, the measure of a system's thermal energy per unit temperature that is unavailable for doing useful work. Because work is obtained from ordered molecular motion, the amount of entropy is also a measure of the molecular disorder, or randomness, of a system. The concept of entropy provides deep insight into the direction of spontaneous change for many everyday phenomena. Its introduction by the German physicist Rudolf Clausius in 1850 is a highlight of 19th-century physics.

The idea of entropy provides a mathematical way to encode the intuitive notion of which processes are impossible, even though they would not violate the fundamental law of conservation of energy. For example, a block of ice placed on a hot stove surely melts, while the stove grows cooler. Such a process is called irreversible because no slight change will cause the melted water to turn back into ice while the stove grows hotter. In contrast, a block of ice placed in an ice-water bath will either thaw a little more or freeze a little more, depending on whether a small amount of heat is added to or subtracted from the system. Such a process is reversible because only an infinitesimal amount of heat is needed to change its direction from progressive freezing to progressive thawing. Similarly, compressed gas confined in a cylinder could either expand freely into the atmosphere if a valve were opened (an irreversible process), or it could do useful work by pushing a moveable piston against the force needed to confine the gas. The latter process is reversible because only a slight increase in the restraining force could reverse the direction of the process from expansion to compression. For reversible processes the system is in equilibrium with its environment, while for irreversible processes it is not.

To provide a quantitative measure for the direction of spontaneous change, Clausius introduced the concept of entropy as a precise way of expressing the second law of thermodynamics. The Clausius form of the second law states that spontaneous change for an irreversible process in an isolated system (that is, one that does not exchange heat or work with its surroundings) always proceeds in the direction of increasing entropy. For example, the block of ice and the stove constitute two parts of an isolated system for which total entropy increases as the ice melts.

For a Carnot cycle, and hence for any reversible processes, $Q_c/Q_h = T_c/T_h$. Rearranging terms yields $\frac{Q_c}{T_c} = \frac{Q_h}{T_h}$ for any reversible process. Q_c and Q_h are absolute values of the heat transfer at temperatures T_c and T_h, respectively. This ratio of Q/T is defined to be the change in entropy ΔS for a reversible process,

$$\Delta S = \left(\frac{Q}{T}\right) \text{rev,}$$

where Q is the heat transfer, which is positive for heat transfer into and negative for heat transfer out of, and T is the absolute temperature at which the reversible process takes place. The SI unit for entropy is joules per kelvin (J/K). If temperature changes during the process, then it is usually a good approximation (for small changes in temperature) to take T to be the average temperature, avoiding the need to use integral calculus to find ΔS.

The definition of ΔS is strictly valid only for reversible processes, such as used in a Carnot engine. However, we can find ΔS precisely even for real, irreversible processes. The reason is that the entropy S of a system, like internal energy U, depends only on the state of the system and not how it reached that condition. Entropy is a property of state. Thus the change in entropy ΔS of a system between state one and state two is the same no matter how the change occurs. We just need to find or imagine a reversible process that takes us from state one to state two and calculate ΔS for that process. That will be the change in entropy for any process going from state one to state two.

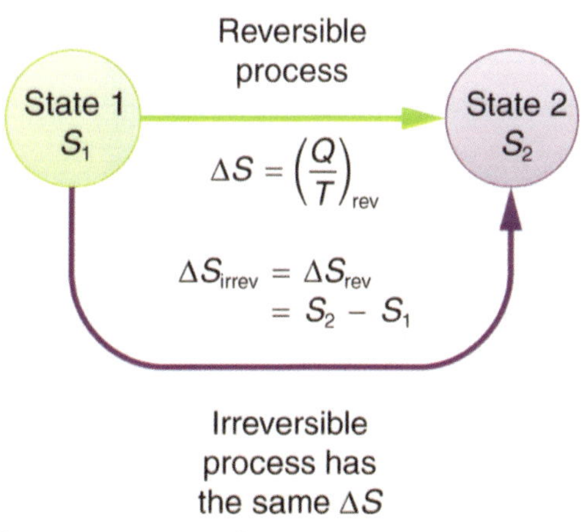

Change in Entropy: When a system goes from state one to state two, its entropy changes by the same amount ΔS, whether a hypothetical reversible path is followed or a real irreversible path is taken

Example

Now let us take a look at the change in entropy of a Carnot engine and its heat reservoirs for one full cycle. The hot reservoir has a loss of entropy $\Delta Sh=-Q_h/T_h$, because heat transfer occurs out of it (remember that when heat transfers out, then Q has a negative sign). The cold reservoir has a gain of entropy $\Delta S_c=Q_c/T_c$, because heat transfer occurs into it. (We assume the reservoirs are sufficiently large that their temperatures are constant.) So the total change in entropy is, PV Diagram for a Carnot Cycle: PV diagram for a Carnot cycle, employing only reversible isothermal and adiabatic processes. Heat transfer Qh occurs into the working substance during the isothermal path AB, which takes place at constant temperature Th. Heat transfer Qc occurs out of the working substance during the isothermal path CD, which takes place at constant temperature Tc. The net work output W equals the area inside the path ABCDA. Also shown is a schematic of a Carnot engine operating between hot and cold reservoirs at temperatures Th and Tc.

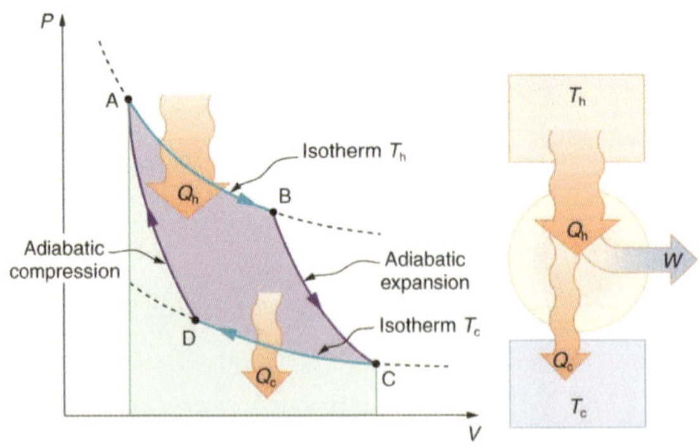

$$\Delta S_{tot} = \Delta S_h + \Delta S_c$$

Thus, since we know that $Q_h/T_h=Q_c/T_c$ for a Carnot engine,

$$\Delta S_{tot} = -\frac{Q_h}{T_h} + \frac{Q_c}{T_c} = 0$$

This result, which has general validity, means that the total change in entropy for a system in any reversible process is zero.

Stastical Interpretation of Entropy

According the second law of thermodynamics, disorder is vastly more likely than order.

The various ways of formulating the second law of thermodynamics tell what happens rather than why it happens. Why should heat transfer occur only from hot to cold? Why should the universe become increasingly disorderly? The answer is that it is a matter of overwhelming probability. Disorder is simply vastly more likely than order. To illustrate this fact, we will examine some random processes, starting with coin tosses.

Coin Tosses

What are the possible outcomes of tossing 5 coins? Each coin can land either heads or tails. On the large scale, we are concerned only with the total heads and tails and not with the order in which heads and tails appear. The following table shows all possibilities along with numbers of possible configurations (or microstate; a detailed description of every element of a system). For example, 4 heads and 1 tail instance may occur on 5 different configurations, with any one of the 5 coins showing tail and all the rest heads. (HHHHT, HHHTH, HHTHH, HTHHH, THHHH)

- 5 heads, 0 tails: 1 microstate
- 4 heads, 1 tail: 5 microstates
- 3 heads, 2 tails: 10 microstates
- 2 heads, 3 tails: 10 microstates
- 1 head, 4 tails: 5 microstates
- 0 head, 5 tails: 1 microstate.

Note that all of these conclusions are based on the crucial assumption that each microstate is equally probable. Otherwise, the analysis will be erroneous.

The two most orderly possibilities are 5 heads or 5 tails. (They are more structured than the others.) They are also the least likely, only 2 out of 32 possibilities. The most disorderly possibilities are 3 heads and 2 tails and its reverse. (They are the least structured.) The most disorderly possibilities are also the most likely, with 20 out of 32 possibilities for the 3 heads and 2 tails and its reverse. If we start with an orderly array like 5 heads and toss the coins, it is very likely that we will get a less orderly array as a result, since 30 out of the 32 possibilities are less orderly. So even if you start with an orderly state, there is a strong tendency to go from order to disorder, from low entropy to high entropy. The reverse can happen, but it is unlikely.

This result becomes dramatic for larger systems. Consider what happens if you have 100 coins instead of just 5. The most orderly arrangements (most structured) are 100 heads or 100 tails. The least orderly (least structured) is that of 50 heads and 50 tails. There is only 1 way (1 microstate) to get the most orderly arrangement of 100 heads. The total number of different ways 100 coins can be tossed—is an impressively large 1.27×10^{30}. Now, if we start with an orderly macrostate like 100 heads and toss the coins, there is a virtual certainty that we will get a less orderly macrostate. If you tossed the coins once each second, you could expect to get either 100 heads or 100 tails once in 2×10^{22} years! In contrast, there is an 8% chance of getting 50 heads, a 73% chance of getting from 45 to 55 heads, and a 96% chance of getting from 40 to 60 heads. Disorder is highly likely.

Real Gas

The fantastic growth in the odds favoring disorder that we see in going from 5 to 100 coins continues as the number of entities in the system increases. In a volume of 1 m^3, roughly 10^{23} molecules (or the order of magnitude of Avogadro's number) are present in a gas. The most likely conditions (or macrostate) for the gas are those we see all the time—a random distribution of atoms in space with a Maxwell-Boltzmann distribution of speeds in random directions, as predicted by kinetic theory as shown in (a). This is the most disorderly and least structured condition we can imagine.

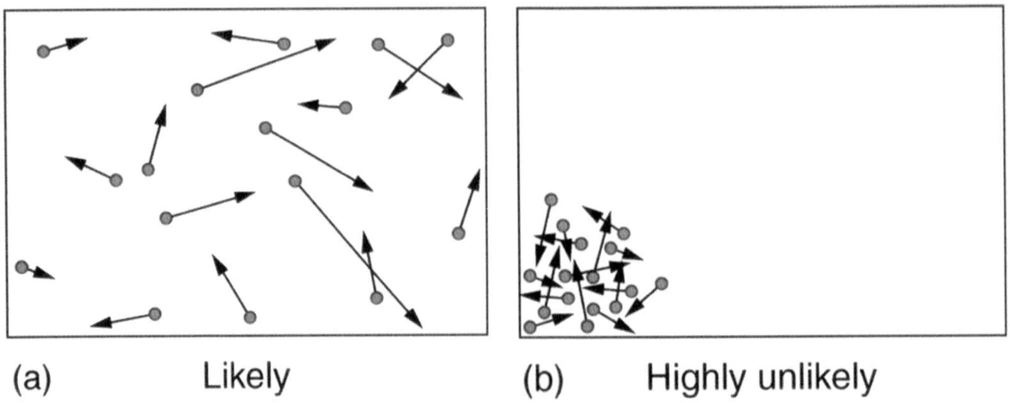

Kinetic Theory: (a) The ordinary state of gas in a container is a disorderly, random distribution of atoms or molecules with a Maxwell-Boltzmann distribution of speeds. It is so unlikely that these atoms or molecules would ever end up in one corner of the container that it might as well be impossible. (b) With energy transfer, the gas can be forced into one corner and its entropy greatly reduced. But left alone, it will spontaneously increase its entropy and return to the normal conditions, because they are immensely more likely.

In contrast, one type of very orderly and structured macrostate has all of the atoms in one corner of a container with identical velocities. There are very few ways to accomplish this (very few microstates corresponding to it), and so it is exceedingly unlikely ever to occur. Indeed, it is so unlikely that we have a law saying that it is impossible, which has never been observed to be violated—the second law of thermodynamics.

Order to Disorder

Entropy is a measure of disorder, so increased entropy means more disorder in the system.

Entropy is a measure of disorder. This notion was initially postulated by Ludwig Boltzmann in the 1800s. For example, melting a block of ice means taking a highly structured and orderly system of water molecules and converting it into a disorderly liquid in which molecules have no fixed positions. There is a large increase in entropy in the process.

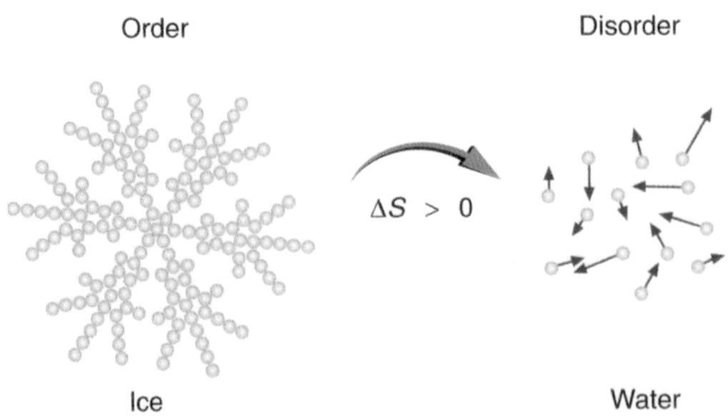

Entropy of Ice: When ice melts, it becomes more disordered and less structured. The systematic arrangement of molecules in a crystal structure is replaced by a more random and less orderly movement of molecules without fixed locations or orientations. Its entropy increases because heat transfer occurs into it. Entropy is a measure of disorder

Example

As an example, suppose we mix equal masses of water originally at two different temperatures, say 20.0° C and 40.0° C. The result is water at an intermediate temperature of 30.0° C. Three outcomes have resulted:

1. Entropy has increased.
2. Some energy has become unavailable to do work.
3. The system has become less orderly.

Entropy, Energy and Disorder

Let us think about each of the results. First, entropy has increased for the same reason that it did in the example above. Mixing the two bodies of water has the same effect as heat transfer from the hot one and the same heat transfer into the cold one. The mixing decreases the entropy of the hot water but increases the entropy of the cold water by a greater amount, producing an overall increase in entropy.

Second, once the two masses of water are mixed, there is only one temperature—you cannot run a heat engine with them. The energy that could have been used to run a heat engine is now unavailable to do work.

Third, the mixture is less orderly, or to use another term, less structured. Rather than having two masses at different temperatures and with different distributions of molecular speeds, we now have a single mass with a uniform temperature.

These three results—entropy, unavailability of energy, and disorder—are not only related but are in fact essentially equivalent.

Heat Death

The entropy of the universe is constantly increasing and is destined for thermodynamic equilibrium, called the heat death of the universe.

In the early, energetic universe, all matter and energy were easily interchangeable and identical in nature. Gravity played a vital role in the young universe. Although it may have seemed disorderly, there was enormous potential energy available to do work—all the future energy in the universe.

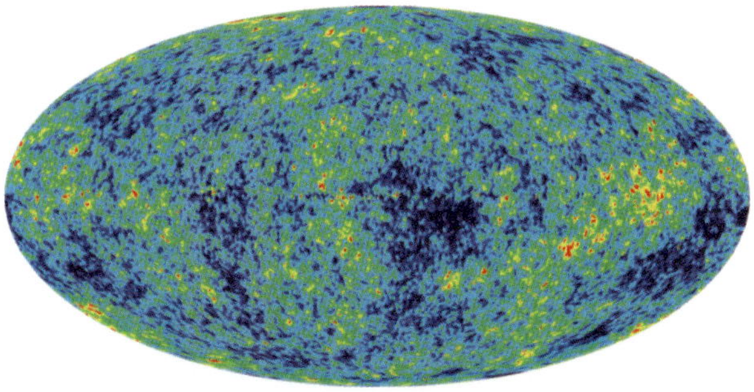

Infant Universe: The image of an infant universe reveals temperature fluctuations (shown as color differences) that correspond to the seeds that grew to become the galaxies

As the universe matured, temperature differences arose, which created more opportunity for work. Stars are hotter than planets, for example, which are warmer than icy asteroids, which are warmer still than the vacuum of the space between them. Most of these are cooling down from their usually violent births, at which time they were provided with energy of their own—nuclear energy in the case of stars, volcanic energy on Earth and other planets, and so on. Without additional energy input, however, their days are numbered.

As entropy increases, less and less energy in the universe is available to do work. On Earth, we still have great stores of energy such as fossil and nuclear fuels; large-scale temperature differences, which can provide wind energy; geothermal energies due to differences in temperature in Earth's layers; and tidal energies owing to our abundance of liquid water. As these are used, a certain fraction of the energy they contain can never be converted into doing work. Eventually, all fuels will be exhausted, all temperatures will equalize, and it will be impossible for heat engines to function, or for work to be done.

Since the universe is a closed system, the entropy of the universe is constantly increasing, and so the availability of energy to do work is constantly decreasing. Eventually, when all stars have died, all forms of potential energy have been utilized, and all temperatures have equalized (depending on the mass of the universe, either at a very high temperature following a universal contraction, or a very low one, just before all activity ceases) there will be no possibility of doing work.

Either way, the universe is destined for thermodynamic equilibrium—maximum entropy. This is often called the heat death of the universe, and will mean the end of all activity. However, whether the universe contracts and heats up, or continues to expand and cools down, the end is not near. Calculations of black holes suggest that entropy can easily continue for at least 10^{100} years.

Light

Light is the common name for electromagnetic radiation with wavelengths of 400-750 nm. This term is sometimes expanded to include ultraviolet and infrared radiation. The human eye is capable of detecting light, which distinguishes it from other regions of the electromagnetic spectrum (despite having shared properties with light, most notably speed).

Both the electric field component and the magnetic field component of electromagnetic radiation travel in space, much like a wave travels in water. In fact, these two components are perpendicular to each other, and are both also at a right angle to the direction of propagation of the wave (*e.g.* if the wave travels along a z-axis, then the electric field might be on the x-axis and the magnetic field on the y-axis). The speed of travel of the wave is constant, equal to about 300,000 kilometers per second in vacuum. According to Einstein's *theory of relativity*, this speed is unique in that it is not affected by the speed of the observer.

Light obviously has a crucial role in our lives; allowing us to see, delivering us the sun's radiation, warming the surface of the Earth, and more. Light is characterized by several very interesting properties, the observation of which has led to the development of the most prominent theories in contemporary physics – the *theory of relativity* and *quantum mechanics*.

Properties of Light

1. Speed of Light : Light travels in the form of straight line. Light travels with constant speed in the vacuum and this constant speed is equal to the $c = 2.9979245\times\times10^{8}$ m s^{-1}. This is approximately taken as $3\times\times10^{8}$ m s^{-1}.

2. Reflection: The basic is meaning of a reflection is something returned in response". When the light reflected from the medium or the surface, the angle of reflection depends only on the angle of incidence to the surface. According to the law of reflection, the angle of reflected wave reflected from the surface is always equal to the angle of incidence.

3. Superposition: If in the space multiple light waves are present then at a particular point in the space the resultant electric field would be the vector sum of the all the electric fields present at that point. This is the law of superposition. As light consists of electric and magnetic field, the law of superposition is followed by both electric and magnetic fields and we get the resultant magnetic and electric fields at that point.

4. Refraction: When light passes through a substance or medium, light gets bend on the basis of the wavelength or frequency. This phenomenon is termed as the Refraction. It follows Snells Law of Refraction.

5. Light Transmittance : When light falls on a substance, some of the light energy gets refracted, some gets refracted, and some gets absorbed by the substance. Light transmittance is defined as the fraction of incident light on the substance which passes through the substance at a particular frequency or wavelength. Light transmittance is defined as the ratio of the intensity of the Incident light on the substance to the intensity of light which passes through the substance or comes out from the substance.

6. Wavelength: Wavelength of Light is defined as the distance between the two consecutive crests or between two consecutive troughs. Wavelength of the light ranges from 380nm to 750nm. In the electromagnetic spectrum, Infrared rays has higher wavelength than the visible rays and the ultraviolet rays has smaller wavelength as compared to the light.

7. Frequency: Frequency of light is defined as the number of crests which passes through a particular point in a second. It is represented in hertz. Frequency of waves is inversely proportional to their wavelength, which means that higher the wavelength, lower is the frequency and vice versa. The frequency range of the light is between 400 to 790 THz.

8. Colors in light: Light consist of different types of colors. These colors are differentiated on the basis of their wavelengths in the visible spectrum . For example, when visible white light is made to pass through the prism, the different colors present in the visible white light bend at different angles depending on their wavelengths and as a result different colors of light are observed. The colors present in the light are Red, orange, yellow, green, and cyan, blue, violet.

Different properties of light like reflection, refraction, scattering, absorption are shown in the figure below.

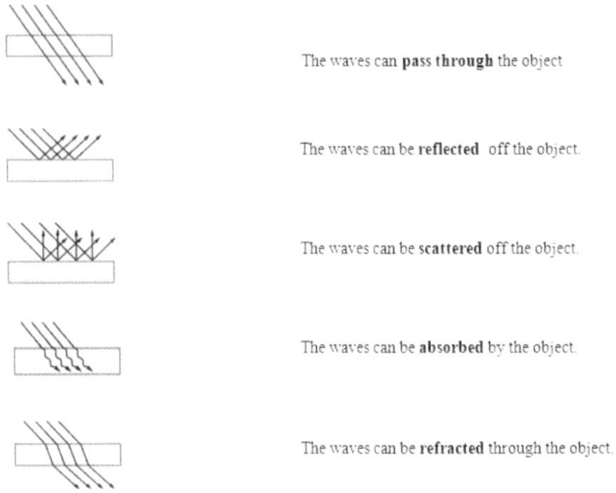

The Duality of Light

The concept of the duality of matter stands at the heart of quantum theory, and it is well established today that matter can possess properties of both waves and particles. With respect to light, this has been recognized for many years, and even during the time of Newton it was realized that light can be described using either linear optics (like a beam of particles), and also using wave-related phenomena, such as interference and diffraction.

Today we can describe light at the macroscopic level using Maxwell's wave equations which describes the propagation of the electromagnetic field in space, while at the microscopic level we know that light is composed of energy quanta in the form of photons.

Each photon's energy is determined solely by its frequency/wavelength (*i.e.*color). This was first speculated by Max Planck in his theoretical studies regarding the electromagnetic spectrum emitted by a "black body", thus laying the most important cornerstone to the development of quantum mechanics.

The Polarity of Light

Another interesting property, which can also be associated with electrons, is the polarity of light.

Light consists of electric field and magnetic field components, both propagating in space in the form of a wave. Let us distinguish between two different situations. In the first one, the direction of the electric field (and hence of the magnetic field) is constant. This is called linear polarity. In the second situation, termed circular polarity, the electric field rotates constantly around the axis along which the light wave propagates. In this case, if the wave travels along the z-axis, the electric field rotates continuously between the x- and y- axes.

This feature of light can be exploited in the making of polarizers, which are devices that block out light waves of a specific linear polarity. The light coming from the sun or from incandescent light bulbs is polarized in a completely random manner. By passing it through a polarizer, one can reduce its intensity (as is done in sunglasses) and obtain polarized light.

Polarity can also be discussed with respect to quantum theory. Since photons possess a spin that can take on one of two possible values (one or negative one), we can make a polarizer that would completely block photons with a spin of negative one and let through only photons with a spin of one. In fact, this property of photons is utilized today as a means to encrypt information. This type of encryption is extremely difficult to decipher without being discovered.

This type of quantum polarity can be described for electric current much like it is described for light. Like photons, electrons also possess a spin, which can take on a value of either half or negative half. Hence, one can think of a similar use of electron spins in order to produce a polarized electric current. However, the implementation of this is much more complicated.

The Interaction of Light and Matter

One of the most interesting questions in the field is how light and matter interact with each other. For example, why are some materials transparent while others are not? Can light change the properties of materials? And so on.

Following the suggestion that light is composed of energy quanta in the form of photons, it was Albert Einstein who proposed a mechanism by which photons that hit a material can be absorbed by it and excite its electrons to a higher energy state, but only if the photons possess the required amount of energy. If not, no matter how many photons hit the material, they would not be able to excite the electrons and be absorbed. It was this *photoelectric effect*that endowed Einstein with a Nobel Prize in Physics in 1921. When this effect was later demonstrated experimentally, it provided a practical proof for the very existence of photons.

The photoelectric effect was an important milestone in our understanding of the interaction between light and matter. A photon can be absorbed only if it can transfer all its energy to the material, exciting its electrons from their ground state to a higher energy level. In doing so the photon is absorbed in the material, which is then said to be opaque to light of the color of the photon. If the photon does not get absorbed, it simply passes through the material.

Since a photon excites electrons when it gets absorbed, it can change some of the properties of the material. For example, a photon can transfer electrons between different energy levels, thus transforming the material from a non-conductive to a conductive state. This phenomenon is exploited in devices such as night vision instruments, photoelectric cells and more.

Einstein's suggestion that photons get absorbed and emitted when electrons are moving between different energy levels led to the realization that this process can be controlled and exploited to produce a unique light beam, the laser. This invention brought about a true technological revolution.

Light as Energy

Light shows both types of behaviors, particle nature as well as the wave nature. According to the particle nature of the light, light consists of particles termed as the photons.

Energy of the light is given by the formula,

$$E = hf$$

Above formulae can also be given as:

We know that the speed of the light can be given as $f = \frac{c}{\lambda}$,

Therefore,

$$E = \frac{hc}{\lambda}$$

Where,

E = Energy of the light wave,

h = Planck's constant,

λ= wavelength of the wave

c = speed of the light

As light consist of photons, it exerts pressure or force on the particles which comes in its path. For example ultraviolet rays are used in the photoelectric effect to eject the electrons from the substances. When rays fall on the substances, photons transfer their energy to the electrons; electrons gain energy and come out from their orbits. Pressure exerted by the light is given as the Power of light divided by the speed of the light.

Light as a energy is also used to heat the substances : When an electron in the excited state in the higher orbit comes to the lower orbit energy is emitted in the form of photons. Due to the transition of electron from the higher orbit to the lower orbit, one photon energy is emitted.

We use light energy in our daily lives to see the world around us. For example incandescent lamps emit light energy, when electricity is given to them. Here in case of incandescent lamps also due to electric energy the electrons of incandescent lamps jumps from the lower energy state to the higher energy state and when they come back to the lower orbit they emit light energy or photons.

Visible Light

Visible Light which is also known as visible spectrum or simply light, the visible spectrum is a portion of the electromagnetic spectrum that the human eye can perceive.

Photometry

Photometry is the astronomical measurement of brightness or intensity and color. It measured as the amount of light energy that strikes a certain surface area on the earth over a certain period of time.

Monochromatic Light

Monochromatic light is the light made of only one color having a single wavelength. An emission lamp is a common source of monochromatic light. For example a mercury or low-pressure sodium streetlight, filtered to remove unwanted emission lines.

Particle

A particle is a fundamental concept in physical sciences on which scientific models of large masses or collection of objects are modeled. It is a localized object to which is attributed the physical and chemical parameters of mass, volume and density. They range in size from subatomic particles such as the electron to microscopic particles such as atoms and molecules, to macroscopic sizes like those of sand, powders or granular materials.

The modeling of natural phenomena is complex and involves intricate computation. The simplest treatment of such phenomena is possible by making simplifying assumptions regarding the processes involved. An example of this is the motion of an object such as a ball that is modeled as a rigid sphere, with negligible friction in air, buoyancy and rotation. This is the classical point sphere treatment. For a system of large number of particles, the physics moves into the realm of statistical mechanics.

Depending on the size of the particles studied, three classes of particles can be distinguished, such as microscopic, macroscopic and subatomic particles. Macroscopic particles refer to particles that are larger than atoms and molecules such as dust, sand or objects as massive as stars or planets. Microscopic particles have sizes between those between atoms and molecules. These can include sub-atomic particles such as protons, neutrons and electrons or nanoparticles, carbon dioxide molecule, water molecule, colloidal particles, etc.

Wave

Wave motion transfers energy from one point to another, usually without permanent displacement of the particles of the medium.

Vibrations and waves are extremely important phenomena in physics. In nature, oscillations are found everywhere. From the jiggling of atoms to the large oscillations of sea waves, we find examples of vibrations in almost every physical system. In physics a wave can be thought of as a disturbance or oscillation that travels through space-time, accompanied by a transfer of energy. Wave motion transfers energy from one point to another, often with no permanent displacement of the particles of the medium —that is, with little or no associated mass transport. They consist, instead, of oscillations or vibrations around almost fixed locations.

The emphasis of the last point highlights an important misconception of waves. Waves transfer energy not mass. An easy way to see this is to imagine a floating ball a few yards out to sea. As the waves propagate (i.e., travel) towards the shore, the ball will not come towards the shore. It may come to shore eventually due to the tides, current or wind, but the waves themselves will not carry the ball with them. A wave only moves mass perpendicular to the direction of propagation—in this case up and down.

A wave can be transverse or longitudinal depending on the direction of its oscillation. Transverse waves occur when a disturbance causes oscillations perpendicular (at right angles) to the propagation (the direction of energy transfer). Longitudinal waves occur when the oscillations are parallel to the direction of propagation. While mechanical waves can be both transverse and longitudinal, all electromagnetic waves are transverse. Sound, for example, is a longitudinal wave.

The description of waves is closely related to their physical origin for each specific instance of a wave process. For example, acoustics is distinguished from optics in that sound waves are related to a mechanical rather than an electromagnetic (light) wave transfer caused by vibration. Therefore, concepts such as mass, momentum, inertia or elasticity become crucial in describing acoustic (as distinct from optic) wave processes. This difference in origin introduces certain wave characteristics particular to the properties of the medium involved. In this chapter we will closely examine the difference between longitudinal and transverse waves along with some of the properties they possess. We will also learn how waves are fundamental in describing motion of many applicable physical systems.

Transverse Waves

Transverse waves propagate through media with a speed →vwv→w orthogonally to the direction of energy transfer.

A transverse wave is a moving wave that consists of oscillations occurring perpendicular (or right angled) to the direction of energy transfer. If a transverse wave is moving in the positive *x*-direction, its oscillations are in up and down directions that lie in the *y*–*z* plane. Light is an example of a transverse wave. For transverse waves in matter, the displacement of the medium is perpendicular to the direction of propagation of the wave. A ripple on a pond and a wave on a string are easily visualized transverse waves.

Transverse waves are waves that are oscillating perpendicularly to the direction of propagation. If you anchor one end of a ribbon or string and hold the other end in your hand, you can create transverse waves by moving your hand up and down. Notice though, that you can also launch waves by moving your hand side-to-side. This is an important point. There are two independent directions in which wave motion can occur. In this case, these are the *y* and *z* directions mentioned above. depicts the motion of a transverse wave. Here we observe that the wave is moving in *t* and oscillating in the *x*-*y* plane. A wave can be thought as comprising many particles which oscillate up and down. In the figure we observe this motion to be in *x*-*y* plane. As time passes the oscillations are separated by units of time. The result of this separation is the sine curve we expect when we plot position versus time.

When a wave travels through a medium–i.e., air, water, etc., or the standard reference medium (vacuum)–it does so at a given speed: this is called the speed of propagation. The speed at which the wave propagates is denoted and can be found using the following formula:

$$v = f\lambda$$

where v is the speed of the wave, f is the frequency , and is the wavelength. The wavelength spans crest to crest while the amplitude is 1/2 the total distance from crest to trough. Transverse waves have their applications in many areas of physics. Examples of transverse waves include seismic S (secondary) waves, and the motion of the electric (E) and magnetic (M) fields in an electromagnetic plane waves, which both oscillate perpendicularly to each other as well as to the direction of energy transfer. Therefore an electromagnetic wave consists of two transverse waves, visible light being an example of an electromagnetic wave.

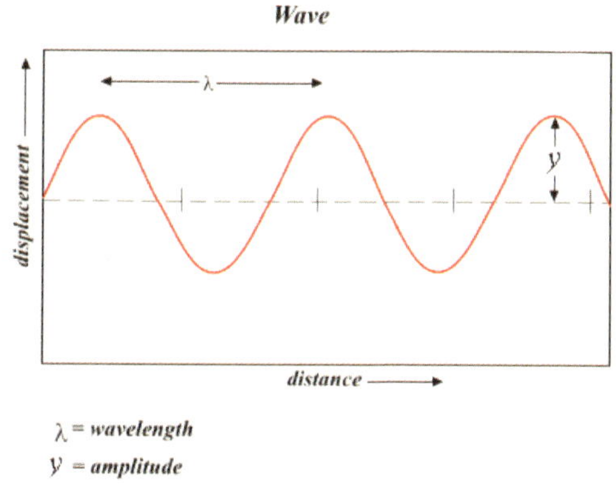

Wavelength and Amplitude: The wavelength is the distance between adjacent crests. The amplitude is the 1/2 the distance from crest to trough

Longitudinal Waves

Longitudinal waves, sometimes called compression waves, oscillate in the direction of propagation.

Longitudinal waves have the same direction of vibration as their direction of travel. This means that the movement of the medium is in the same direction as the motion of the wave. Some longitudinal waves are also called compressional waves or compression waves. An easy experiment for observing longitudinal waves involves taking a Slinky and holding both ends. After compressing and releasing one end of the Slinky (while still holding onto the end), a pulse of more concentrated coils will travel to the end of the Slinky.

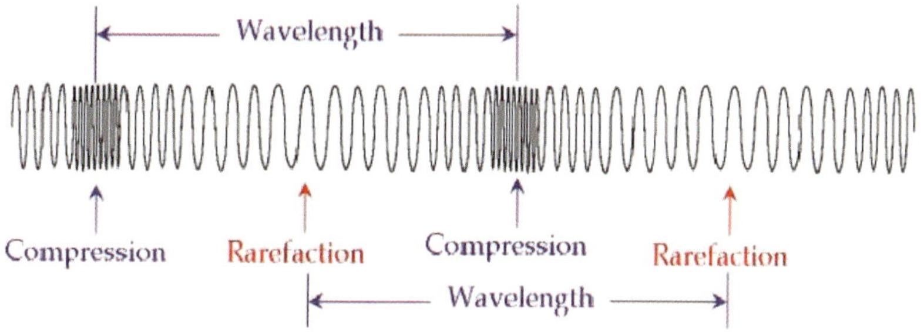

Longitudinal Waves: A compressed Slinky is an example of a longitudinal wave. The wave propagates in the same direction of oscillation

Like transverse waves, longitudinal waves do not displace mass. The difference is that each particle which makes up the medium through which a longitudinal wave propagates oscillates along the axis of propagation. In the example of the Slinky, each coil will oscillate at a point but will not travel the length of the Slinky. It is important to remember that energy, in this case in the form of a pulse, is being transmitted and not the displaced mass.

Longitudinal waves can sometimes also be conceptualized as pressure waves. The most common pressure wave is the sound wave. Sound waves are created by the compression of a medium, usually air. Longitudinal sound waves are waves of alternating pressure deviations from the equilibrium pressure, causing local regions of compression and rarefaction. Matter in the medium is periodically displaced by a sound wave, and thus oscillates. When people make a sound, whether it is through speaking or hitting something, they are compressing the air particles to some significant amount. By doing so, they create transverse waves. When people hear sounds, their ears are sensitive to the pressure differences and interpret the waves as different tones.

Water Waves

Water waves can be commonly observed in daily life, and comprise both transverse and longitudinal wave motion.

Water waves, which can be commonly observed in our daily lives, are of specific interest to physicists. Describing detailed fluid dynamics in water waves is beyond the scope of introductory physics courses. Although we often observe water wave propagating in 2D, in this atom we will limit our discussion to 1D propagation.

Water waves: Surface waves in water

The uniqueness of water waves is found in the observation that they comprise both transverse and longitudinal wave motion. As a result, the particles composing the wave move in clockwise circular motion, as seen in. Oscillatory motion is highest at the surface and diminishes exponentially with depth. Waves are generated by wind passing over the surface of the sea. As long as the waves propagate slower than the wind speed just above the waves, there is an energy transfer from the wind to the waves. Both air pressure differences between the upwind and the lee side of a wave crest, as well as friction on the water surface by the wind (making the water to go into the shear stress), contribute to the growth of the waves.

In the case of monochromatic linear plane waves in deep water, particles near the surface move in circular paths, creating a combination of longitudinal (back and forth) and transverse (up and down) wave motions. When waves propagate in shallow water (where the depth is less than half the wavelength), the particle trajectories are compressed into ellipses. As the wave amplitude (height) increases, the particle paths no longer form closed orbits; rather, after the passage of each crest, particles are displaced slightly from their previous positions, a phenomenon known as Stokes drift.

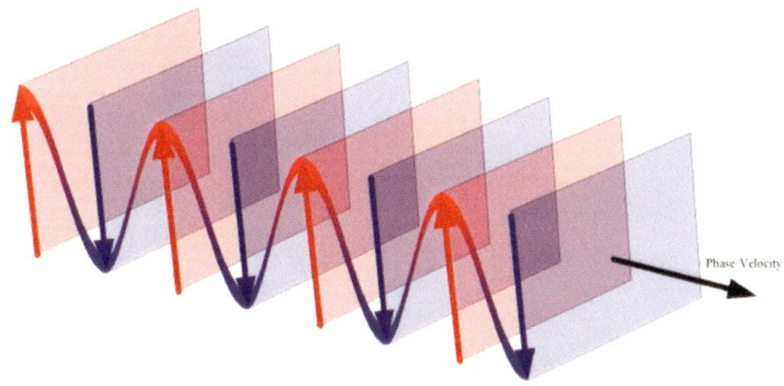

Plane wave: We see a wave propagating in the direction of the phase velocity.
The wave can be thought to be made up of planes orthogonal to the direction of the phase velocity

Since water waves transport energy, attempts to generate power from them have been made by utilizing the physical motion of such waves. Although larger waves are more powerful, wave power is also determined by wave speed, wavelength, and water density. Deep water corresponds with a water depth larger than half the wavelength, as is a common case in the sea and ocean. In deep water, longer-period waves propagate faster and transport their energy faster. The deep-water group velocity is half the phase velocity. In shallow water for wavelengths larger than about twenty times the water depth (as often found near the coast), the group velocity is equal to the phase velocity. These methods have proven viable in some cases but do not provide a fully sustainable form of renewable energy to date.

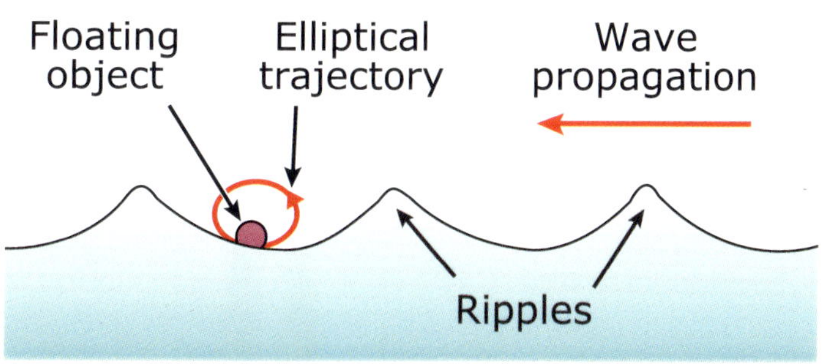

Water waves: The motion water waves causes particles to follow clockwise circular motion.
This is a result of the wave having both transverse and longitudinal properties

Wavelength, Freqency in Relation to Speed

Waves are defined by its frequency, wavelength, and amplitude among others. They also have two kinds of velocity: phase and group velocity.

Characteristics of Waves

Waves have certain characteristic properties which are observable at first notice. The first property to note is the amplitude. The amplitude is half of the distance measured from crest to trough. We also observe the wavelength, which is the spatial period of the wave (e.g. from crest to crest or trough to trough). We denote the wavelength by the Greek letter λλ.

The frequency of a wave is the number of cycles per unit time — one can think of it as the number of crests which pass a fixed point per unit time. Mathematically, we make the observation that,

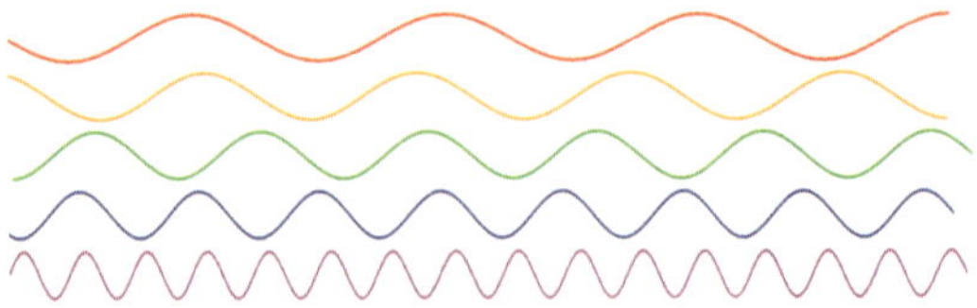

Frequencies of different sine waves.: The red wave has a low frequency sine there is very little repetition of cycles. Conversely we say that the purple wave has a high frequency. Note that time increases along the horizontal

$$f = \frac{1}{T}$$

where T is the period of oscillation. Frequency and wavelength can also be related-* with respects to a "speed" of a wave. In fact,

$$v = f\lambda$$

where v is called the wave speed, or more commonly,the phase velocity, the rate at which the phase of the wave propagates in space. This is the velocity at which the phase of any one frequency component of the wave travels. For such a component, any given phase of the wave (for example, the crest) will appear to travel at the phase velocity.

Finally, the group velocity of a wave is the velocity with which the overall shape of the waves' amplitudes — known as the modulation or envelope of the wave — propagates through space. In, one may see that the overall shape (or "envelope") propagates to the right, while the phase velocity is negative.

Energy Transportation

Waves transfer energy which can be used to do work.

Energy transportion is essential to waves. It is a common misconception that waves move mass. Waves carry energy along an axis defined to be the direction of propagation. One easy example is to imagine that you are standing in the surf and you are hit by a significantly large wave, and once you are hit you are displaced (unless you hold firmly to your ground!). In this sense the wave has done work (it applied a force over a distance). Since work is done over time, the energy carried by a wave can be used to generate power.

Water Wave: Waves that are more massive or have a greater velocity transport more energy

Similarly we find that electromagnetic waves carry energy. Electromagnetic radiation (EMR) carries energy—sometimes called radiant energy—through space continuously away from the source (this is not true of the near-field part of the EM field). Electromagnetic waves can be imagined as a self-propagating transverse oscillating wave of electric and magnetic fields. EMR also carries both momentum and angular momentum. These properties may all be imparted to matter with which it interacts (through work). EMR is produced from other types of energy when created, and it is converted to other types of energy when it is destroyed. The photon is the quantum of the electromagnetic interaction, and is the basic "unit" or constituent of all forms of EMR. The quantum nature of light becomes more apparent at high frequencies (or high photon energy). Such photons behave more like particles than lower-frequency photons do.

In general, there is a relation of waves which states that the velocity (v) of a wave is proportional to the frequency (f) times the wavelength (λ):

$$v = f\lambda$$

We also know that classical momentum p is given by p=mv which relates to force via Newton's second law: $F=\frac{dp}{dt}$

EM waves with higher frequencies carry more energy. This is a direct result of the equations above. Since $v \propto f$ we find that higher frequencies imply greater velocity. If velocity is increased then we have greater momentum which implies a greater force (it gets a little bit tricky when we talk about particles moving close to the speed of light, but this observation holds in the classical sense). Since energy is the ability of an object to do work, we find that for W=Fd a greater force correlates to more energy transfer. Again, this is an easy phenomenon to experience empirically; just stand in front of a faster wave and feel the difference.

References

- Smith, Crosbie (1998). The Science of Energy – a Cultural History of Energy Physics in Victorian Britain. The University of Chicago Press. ISBN 0-226-76420-6
- Martyushev, L.M.; Seleznev, V.D. (2006). "Maximum entropy production principle in physics, chemistry and biology". Phys. Rev. 426: 1–45. Bibcode:2006PhR...426....1M. doi:10.1016/j.physrep.2005.12.001
- What-is-space-24870: space.com, Retrieved 18 May 2018
- "E. Noether's Discovery of the Deep Connection Between Symmetries and Conservation Laws". Physics.ucla.edu. 1918-07-16. Archived from the original on 2011-05-14. Retrieved 2010-12-12
- Lofts, G; O'Keeffe D; et al. (2004). "11 — Mechanical Interactions". Jacaranda Physics 1(2 ed.). Milton, Queensland, Australia: John Willey & Sons Australia Ltd. p. 286. ISBN 0-7016-3777-3
- Physics-of-time: exactlywhatistime.com, Retrieved 22 March 2018
- Belkin, A.; et., al. (2015). "Self-Assembled Wiggling Nano-Structures and the Principle of Maximum Entropy Production". Sci. Rep. 5. Bibcode:2015NatSR...5E8323B. doi:10.1038/srep08323
- Bicycle calculator - speed, weight, wattage etc. "Archived copy". Archived from the original on 2009-05-13. Retrieved 2009-05-29.
- Entropy-physics, science: britannica.com, Retrieved 22 March 2018
- Feynman, Richard (1964). The Feynman Lectures on Physics; Volume 1. U.S.A: Addison Wesley. ISBN 0-201-02115-3
- Physics-light, maagarmada: davidson.weizmann.ac.il, Retrieved 29 April 2018

2

Quantum Mechanics

Quantum mechanics is a fundamental theory of physics that is concerned with the study of phenomena at the smallest energy scales of atomic and subatomic particles. The aim of this chapter is to explore the key concepts of quantum mechanics including superposition, tunneling, quantum entanglement, wave function, wave particle duality, uncertainty principle, etc.

Quantum mechanics is the branch of physics relating to the very small. It results in what may appear to be some very strange conclusions about the physical world. At the scale of atoms and electrons, many of the equations of classical mechanics, which describe how things move at everyday sizes and speeds, cease to be useful. In classical mechanics, objects exist in a specific place at a specific time. However, in quantum mechanics, objects instead exist in a haze of probability; they have a certain chance of being at point A, another chance of being at point B and so on.

Three Revolutionary Principles

Quantum mechanics (QM) developed over many decades, beginning as a set of controversial mathematical explanations of experiments that the math of classical mechanics could not explain. It began at the turn of the 20th century, around the same time that Albert Einstein published his theory of relativity, a separate mathematical revolution in physics that describes the motion of things at high speeds. Unlike relativity, however, the origins of QM cannot be attributed to any one scientist. Rather, multiple scientists contributed to a foundation of three revolutionary principles that gradually gained acceptance and experimental verification between 1900 and 1930. They are:

- Quantized properties: Certain properties, such as position, speed and color, can sometimes only occur in specific, set amounts, much like a dial that "clicks" from number to number. This challenged a fundamental assumption of classical mechanics, which said that such properties should exist on a smooth, continuous spectrum. To describe the idea that some properties "clicked" like a dial with specific settings, scientists coined the word "quantized."
- Particles of light: Light can sometimes behave as a particle. This was initially met with harsh criticism, as it ran contrary to 200 years of experiments showing that light behaved as a wave; much like ripples on the surface of a calm lake. Light behaves similarly in that it bounces off walls and bends around corners, and that the crests and troughs of the wave can add up or cancel out. Added wave crests result in brighter light, while waves that cancel out produce darkness. A light source can be thought of as a ball on a stick being rhythmically dipped in the center of a lake. The color emitted corresponds to the distance between the crests, which is determined by the speed of the ball's rhythm.
- Waves of matter: Matter can also behave as a wave. This ran counter to the roughly 30 years of experiments showing that matter (such as electrons) exists as particles.

Mathematical Formulation of Quantum Mechanics

All physical theories are based on fundamental laws formulated in a mathematical framework and on correspondence rules mapping elements of the mathematical theory to physical objects. Compared to classical mechanics, the mathematical framework for quantum mechanics is difficult and the correspondence rules are less intuitive. We start be providing the mathematical basis for the Schr¨odinger representation of quantum mechanics.

Mathematical Prerequisites

A vector space H over the field C of complex numbers, equipped with a scalar product h $\langle \bullet, \bullet \rangle$, which is complete with respect to the norm $\|\psi\| := \sqrt{\langle \psi,\ \psi \rangle}$ induced by the scalar product, is called a Hilbert space.

A Hilbert space H is called separable, if it contains a denumerable complete orthonormal system, i.e. a sequence $\{ui\}_{i\in} IN \subset H$ such that,

$$(i)\ \langle u_{i,} u_j \rangle = \delta_{ij} \qquad \forall i, j \in IN,$$

$$(ii)\ \psi = \sum_{i=1}^{\infty} \langle \psi, u_i \rangle u_i \ \ \forall \psi \in H.$$

A linear map $A : D(A) \subset H \rightarrow H$ is called a linear operator in H. It is bounded if there exists a constant M ≥ 0, such that,

$$\|A\psi\| \le M\|\psi\| \quad \forall\ \psi\ \in H\ .$$

Bounded linear operators can always be extended to H, and therefore for bounded linear operators we shall always consider the situation D(A) = H.

A bounded linear operator is nonnegative, if,

$$\langle A\psi, \psi \rangle \ge 0 \quad \forall \psi \in H.$$

Let A be a nonnegative operator and $\{ui\}_{i\in} IN$ a complete orthonormal set. Then the trace of A is defined by (the finite or infinite value),

$$\mathrm{tr}(A) = \sum_{i=1}^{\infty} \langle Au_i\ ,\ u_i \rangle .$$

A simple computation shows that the value of the trace is independent from the choice of the complete orthonormal system. With a little care this definition can be extended to bounded but not necessarily nonnegative operators. Operators with a finite trace are called trace (class) operators. For a linear operator A, the adjoint operator A^* is defined by,

$$\langle A^* u,\ v \rangle = \langle u,\ Av \rangle\ \forall\ u\ \in\ D(A^*),\ v\ \in\ D(A),$$

where the domain of A^* is defined by,

$$D(A^*) = \{u \in H : \exists w \in H : \langle u, Av\rangle = \langle w, v\rangle \ \forall v \in D(A)\}\ .$$

By the Riesz theorem, $D(A^*) = H$ holds for all bounded operators A. A bounded operator U satisfying $UU^* = U^*U = I$ is called unitary. Under a unitary transformation, the scalar product is invariant:

$$\langle Uu, Uv\rangle = \langle u, v\rangle \ \forall u, v \in H\ .$$

A linear operator A is called self adjoint, iff $A^* = A$ and $D(A^*) = D(A)$ holds. Symmetry, i.e.,

$$\langle Au, v\rangle = \langle u, Av\rangle \ \forall u, v \in D(A),$$

is neccessary for self adjointness. Obviously, for bounded operators it is also sufficient.

The resolvant set of a linear operator A is the set of all $\lambda \in C$ such that the resolvant,

$$R(\lambda, A) := (\lambda I - A)\ -1$$

is a bounded operator. The complement (in C) of the resolvant set is called the spectrum of A and is denoted by $\sigma(A)$. Self adjoint operators A satisfy $\sigma(A) \subset IR$.

The point spectrum $\sigma_p(A) \subset \sigma(A)$ is the set of all eigenvalues, i.e.,

$$\lambda \in \sigma_p(A) \Longleftrightarrow \exists u \in D(A),\ u \neq 0 : Au = \lambda u\ .$$

It has to be noted that in general the inclusion $\sigma_p(A) \subset \sigma(A)$ is strict and that, in particular, a continuous spectrum may occur along with a discrete point spectrum.

For a self adjoint operator A in a finite dimensional Hilbert space (say, dimH = n), there exists an orthonormal basis $\{u_1, \ldots, u_n\}$ of H consisting of eigenvectors of A, i.e., $Au_k = \lambda_k u_k,\ k = 1, \ldots, n.$ Denoting the projection onto the k-th eigenspace of A by $P_k u = \langle u, u_k\rangle u_k$, we have the formula,

$$A = \sum_{k=1}^{n} \lambda_k P_k,$$

which can be written in the form,

$$A\text{ë}= \int_{-\infty}^{\infty} dE(\lambda)\ ,$$

with the spectral family of projections,

$$E(\text{ë}) = \sum_{\lambda_k \leq \lambda} P_k\ .$$

The spectral family satisfies the properties,

$$(i)\ E(\lambda)E(\mu) = E(\lambda)\ for\ \lambda \leq \mu\ ,$$
$$(ii)\ E(-\infty) = 0\ ,\ E(\infty) = I\ ,$$
$$(iii)\ E(\lambda + 0) = E(\lambda).$$

This setting can be transferred to the infinite dimensional case. The spectral theorem asserts that for self adjoint operators in arbitrary Hilbert spaces a spectral family of projections with the above properties exists, such that $E(\lambda) = \sum_{\lambda_k \le \lambda} P_k$. holds. The differential dE(λ) vanishes away from σ(A) and, thus, the integral in $E(\lambda) = \sum_{\lambda_k \le \lambda} P_k$.only contains contributions from the spectral values of A. The spectral theorem provides a framework for applying functions to self adjoint operators:

$$f(A) := \int_{-\infty}^{\infty} f(\lambda) dE(\lambda).$$

The Postulates

One of the experimental observations leading to the development of quantum mechanics was the fact that certain physical quantities, which should take values from a continuum according to classical physics, take only discrete (or quantized) values in certain experiments. This is the case, for example, for the energy emitted by an electron in an atom. Free electrons, on the other hand, have a continuous range of energies. This should be a motivation for the first two postulates:

- Postulate 1: To every quantum mechanical system there corresponds a separable Hilbert space H.
- Postulate 2: Every physical quantity (observable) is represented by a self adjoint operator A in H. The physical quantity can only take values in the spectrum of A.

Another experimental observation shows that the outcome of experiments of quantum mechanical systems cannot be predicted deterministically, but only in a probabilistic sense. The same could be said within the framework of classical mechanics as a result of measurement errors and/or the complexity of a system. In quantum mechanics, however, the probabilistic nature is seen as a basic principle and, therefore, has to be part of every quantum mechanical theory.

The state of a system is therefore always understood as the description of an ensemble of systems with identical experimental conditions. To motivate the representation of states, let us consider a situation with a finite dimensional H and a self adjoint operator A with spectral decomposition. We shall compute the expected value of the observable corresponding to A. By postulate 2, the observable can only take one of the values $\lambda_1, \ldots, \lambda_n$. If the system is in a fixed state, these values are taken with probabilities $p_1, \ldots, p_n \ge 0$ $(p_1 + \cdots + p_n = 1)$. The expected value of A is, thus,

$$\langle A \rangle = \sum_{k=1}^{n} pk\lambda k = \sum_{k=1}^{n} pk \langle Au_k, uk \rangle = tr({}_Q A)$$

with $o_{ij} = \sum_{k=1}^{n} p_k u_{k,i} u_k, j$.

The operator Q is nonnegative and satisfies tr(Q) = 1.

Postulate 3: The state of a quantum mechanical system is represented by a nonnegative trace operator % with tr(Q) = 1, the state operator (or density matrix). The expectation value of an observable A is given by,

$$\langle A \rangle = tr\left({}_Q A\right)$$

The definition of the probabilistic setting is completed by using the spectral family E(λ) corresponding to the observable A. We define the distribution function for A by,

$$prob\left(A \subset (-\infty, \lambda]\right) = tr\left({}_Q E(\lambda)\right).$$

The spectral theorem shows that this is consistent with the formula for the expectation value in postulate 3.

The properties of the state operator immediately imply σ(%) ⊂ [0, 1]. The projection onto a normalized vector $\psi \in H, \ \|\psi\| = 1$,, defines a pure state,

$${}_Q u = \langle u, \psi \rangle \psi.$$

For a pure state, the formula for the distribution function and for the expectation value is simple:

$$prob\left(A \subset (-\infty, \lambda]\right) = \langle E(\lambda)\psi, \psi \rangle, \ \langle A \rangle = \langle A\psi, \psi \rangle.$$

Non-pure states are called mixed states. The spectral theorem applied to state operators has the interpretation that mixed states can be written as convex combinations of pure states (at least in the case of a discrete σ(0) = σp(0)).

Note that for arbitrary real ω, ψ and e $e^{i\omega}\psi$ define the same state. For a pure state, the vector ψ is only unique up to a phase factor $e^{i\omega}\psi$.

Dynamics of a Single Free Particle

If the quantum mechanical system consists of a single particle, we introduce the vector valued position operator $Q = (Q_1, Q_2, Q_3)$, with $\sigma(Q_j) = IR$, j = 1, 2, 3, and the Q_j have common eigenvectors: For every x ∈ IR^3 there is a u(x) ∈ H such that $Q_j u(x) = x_j u(x)$, j = 1, 2, 3. The velocity operator $V = (V_1, V_2, V_3)$ is then defined by the requirement,

$$\frac{d}{dt}\langle Q \rangle = \langle V \rangle.$$

In particular, for a time dependent pure state, given by ψ(t) ∈ H this gives,

$$\langle V\psi, \psi \rangle = \frac{d}{dt}\langle Q\psi, \psi \rangle = \left\langle Q \frac{d\psi}{dt}, \psi \right\rangle + \left\langle Q\psi, \frac{d\psi}{dt} \right\rangle$$

Since H is the generator of the time shift transformation, we have exp (is H) ψ (t) = ψ(t + s). Comparing the generators of both sides (derivatives with respect to s at s = 0) gives,

$$\frac{d\psi}{dt} = iH\psi,$$

and, thus,

$$V = i[Q, H] .$$

We shall need commutation relations between Q and P = (P_1, P_2, P_3). We start by postulating the action of the space shift operator on the eigenvectors of Q:

$$e^{-ia.P} u(x) = u(x + a)$$

with the consequence,

$$e^{ia \cdot P} Q e^{-ia \cdot P} u(x) = (x + a) u(x),$$

Giving,

$$e^{ia \cdot P} Q e^{-ia \cdot P} = Q + aI .$$

Taylor expansion for small a leads to the commutation relations,

$$[Q_j , P_k] = i\delta_{jk} I .$$

As a next step we postulate the following relations between operators:

$$J = Q \times P , H = -\frac{P \cdot P}{2M} ,$$

with J = (J_1, J_2, J_3). These satisfy all the commutation relations stated so far. Substitution of the second relation into $V = i[Q, H]$ .gives,

$$\text{P} = \text{MV} , \text{ and, thus,}$$

$$-H = \frac{1}{2} MV \cdot \text{V} ,$$

$$\text{J} = \text{Q} \times MV .$$

If M where the mass of the particle, these where reasonable definitions of momentum, kinetic energy, and angular momentum. However, comparison of dimensions shows that M has to be measured in $\frac{sec}{cm^2}$. It is plausible that M can be written as,

$$M = \frac{m}{\hbar} ,$$

where m is the mass of the particle and $\hbar$ is a fundamental constant with the dimension of an action (Wirkung). It is called the Planck constant (Plancksches Wirkungsquantum), and its value has been determined experimentally: $\hbar \approx 1.054573 \times 10^{-34}$ joule-seconds.

At this point we conveniently redefine the operators $P \to P/\hbar$, $H \to -H/\hbar$, $\mathbf{J} \to \mathbf{J}/\hbar$, such that the new operators correspond to momentum, kinetic energy, and angular momentum.

Referring to classical mechanics, the Hamiltonian H for a particle moving in a position dependent external field given by the vector potential *A(x)* and a scalar potential *W(x)* is defined as,

$$H = \frac{1}{2}m\big(V - A(Q)\big) \cdot \big(V - A(Q)\big) + W(Q).$$

The Equations of Motion and Conservation Laws

With the Hamiltonian given above, the state vector ψ(t) of a time dependent pure state satisfies the abstract ordinary differential equation,

$$i\hbar\frac{d\psi}{dt} = H\psi .$$

For the corresponding density matrix defined by ${}_Q u = \langle u, \psi\rangle\psi$ a straightforward computation shows,

$$i\hbar\frac{d_Q}{dt} = [H, Q],$$

which will also be assumed to hold for general (mixed) states.

For an observable A the evolution of the expected value is governed by,

$$i\hbar\frac{d}{dt}\langle A\rangle = \langle [A, H]\rangle,$$

with the consequence that the expected values of all observables, which commute with the Hamiltonian, are conserved in time. Examples for the free particle are the identity, P, J, and H, corresponding to conservation of mass, momentum, angular momentum, and energy.

Coordinate Representation and the Schrödinger Equation

For a one particle system we choose the Hilbert space $H = L^2(IR^3)$, i.e. state vectors are complex valued square integrable wave functions ψ(x) satisfying,

$$1 = \|\psi\|^2 = \int_{R^3} |\psi(x)|^2\, dx .$$

The action of the position and translation operators are defined as,

$$(Q\psi)(x) = x\psi(x) \text{ and } \left(e^{ia\cdot P/\hbar}\psi\right)(x) = \psi(x + a).$$

Consequently, the momentum and angular momentum operators are given by,

$$(P\psi)(x) = -i\hbar\nabla\psi(x),\ (J\psi)(x) = -i\hbar x \times \nabla\psi(x),$$

and the Hamiltonian of a particle in an external field by,

$$(H\psi)(x) = \frac{m}{2}\left(\frac{i\hbar}{m}\nabla + A(x)\right)\cdot\left(\frac{i\hbar}{m}\nabla\psi(x) + A(x)\psi(x)\right) + W(x)\psi(x).$$

For vanishing vector potential A, the equation of motion for a time dependent wave vector becomes the Schrödinger equation,

$$i\hbar \frac{\partial \psi}{\partial t} = -\frac{\hbar^2}{2m} \Delta\psi + W\,\psi\,.$$

The density matrix o(x, y, t) of a pure state is,

$$Q(x, y, t) = \psi(x, t)\overline{\psi(y, t)}.$$

For a mixed state its trace also has to be unity:

$$tr(Q) = \int_{R3} Q(x,x,t)\,dx = 1.$$

The density matrix of a time dependent state has to satisfy the von Neum-mann equation,

$$ih\frac{\partial_Q}{\partial t} = -\frac{h^2}{2m}(\Delta_x Q - \Delta y Q) + (W(x) - W(y))Q.$$

Returning to pure states, the probability to find the particle at time t in the domain $\Omega \subset IR^3$ is given by,

$$\int_\Omega |\psi(x, t)|^2\, dx.$$

Computing the rate of change of this quantity, and using the arbitrariness of Ω we derive the conservation equation,

$$\frac{\partial}{\partial t}|\psi|^2 + \nabla \cdot j = 0,$$

with the probability flux,

$$j = \frac{\hbar}{m} Im\left(\overline{\psi}\nabla\psi\right).$$

Let us know check some of the properties stated in the abstract setting of the previous sections. For example, a Galilei transformation $x' = x - vt,\ t' = t$, should leave the Schrödinger equation invariant, if the wave function is multplied by a phase factor: $\psi(x, t) = e^{i\omega(x,t)}\psi'(x',t')$. Substitution in the Schrödinger equation $i\hbar \frac{\partial \psi}{\partial t} = -\frac{\hbar^2}{2m} \Delta\psi + W\,\psi$. shows that the appropriate choice for the phase is $\omega(x, t) = \frac{m}{\hbar} v \cdot (x - vt)$.

Now we turn to the conservation laws. Integration of $\frac{\partial}{\partial t}|\psi|^2 + \nabla \cdot j = 0$, shows that the property $\int_{R^3} |\psi|^2\, dx = 1$ is conserved. The expectation value for the momentum is,

$$\langle P \rangle = \langle -i\hbar\nabla\psi,\, \psi\rangle = \hbar\, Im \int_{R3} \overline{\psi}\nabla\psi\, dx\,.$$

A straightforward computation shows that $\frac{d}{dt}\langle P\rangle = 0$ holds if the potential vanishes, as expected. On the other hand, the energy,

$$\langle H\rangle = \int_{R^3}\left(\frac{\hbar 2}{2m}|\nabla\psi|^2 + W|\psi|^2\right)dx$$

is conserved, when ψ solves $i\hbar\frac{\partial\psi}{\partial t} = -\frac{\hbar^2}{2m}\Delta\psi + W\psi : \frac{d}{dt}\langle H\rangle = 0$

A fundamental consequence of the quantum theory is the Heisenberg uncertainty principle, i.e., the fact that certain pairs of observables cannot be measured simultaneously with arbitrary accuracy. One of these pairs is position and momentum.

The variance ΔA of the observable A is determined by,

$$(\Delta A)^2 = \left\langle (A - \langle A\rangle)^2\right\rangle = \|(A - \langle A\rangle)\psi\|^2$$

The uncertainty principle for position and momentum then takes the form of the inequality,

$$\Delta Q_j \Delta P_j \geq \frac{\hbar}{2},$$

for every pure state. The proof relies on the commutation relation,

$$\frac{i}{\hbar}[P_j, Q_k] = \delta_{jk} I,$$

and on the identities,

$$[A - \langle A\rangle, B - \langle B\rangle] = [A, B], \quad i\langle [A, B]\rangle = 2\, Im\langle A\psi, B\psi\rangle,$$

where the latter holds for symmetric operators A and B:

$$1 = \langle I\rangle = \frac{i}{\hbar}\langle [P_j, Q_j]\rangle = \frac{i}{\hbar}\langle [P_j - \langle P_j\rangle, Q_j - \langle Q_j\rangle]\rangle$$

$$= \frac{2}{\hbar} Im\langle (P_j - \langle P_j\rangle)\psi, (Q_j - \langle Q_j\rangle)\psi\rangle \leq \frac{2}{\hbar}\|(P_j - \langle P_j\rangle)\psi\| \|(Q_j - \langle Q_j\rangle)\psi\|$$

$$= \frac{2}{\hbar}\Delta P_j \Delta Q_j.$$

Simple solutions of the Schrödinger equations can be obtained from eigenfunctions of the Hamiltonian: Let (E, u(x)) denote an eigenvalueeigenfuction-pair, i.e., Hu = Eu. Then,

$$\psi(x, t) = e^{-iEt/\hbar}u(x)$$

is a solution of the Schrödinger equation. Since the time dependence only occurs in a phase factor, all observables of this solution are constant in time. Therefore a solution of this form is called a steady state solution with energy E.

Eigenfunctions are only solutions of Hu = Eu with $u \in L^2(IR^3)$. However, it can be shown that the continuous spectrum can also be recovered by solving this equation, looking for solutions u, not necessarily in L^2 (IR^3), but satisfying certain growth conditions at infinity. These so called spectral functions also correspond to pysically meaningful states of the form above eq.

For example, the functions $u_k(x) = e^{ik\cdot x}$, $k \in R^3$, are spectral functions of the Hamiltonian of a free particle (W = 0) corresponding to the spectral values,

$$E_k = \frac{\hbar^2 |k|^2}{2m}.$$

Tunnelling

The tunnelling effect is one of the striking differences between classical and quantum mechanics. A particle may cross a potential barrier higher than the energy of the particle.

Let us consider a rectangular potential barrier:

$$W(x) = \begin{cases} V_0 & \text{for } 0 \langle x \langle a, \\ 0 & \text{elsewhere.} \end{cases}$$

The one-dimensional steady state Schrödinger equation with energy E is given by,

$$-\frac{\hbar^2}{2m}\frac{d^2\psi}{dx^2} + W\,\psi = E\psi\ ,$$

where we assume $0 < E < V_0$. A classical particle with energy E, travelling towards x = 0 from the left, would be reflected by the barrier. Setting $E = \frac{\hbar^2 k^2}{2m}$, $k > 0$ and $V_0 - E = \frac{\hbar^2\beta^2}{2m}$, solutions of above equations have the form,

$$\psi(x) = \begin{cases} A_1 e^{ikx} + B_1 e^{-ikx}, & x \le 0, \\ Ce^{\beta x} + De^{-\beta x}\ , & 0 \le x \le a\ , \\ A_2 e^{ikx} + B_2 e^{-ikx} & x \ge a\ . \end{cases}$$

The number of free constants is reduced to 2 by requiring ψ to be continuously differentiable.

The probability flux, computed in the three regions, is given by,

$$j = \frac{\hbar}{m}\mathrm{Im}\left(\bar{\psi}\frac{d\psi}{dx}\right) = \frac{\hbar k}{m}\left(|A1|^2 - |B1|^2\right) = \frac{2\hbar\beta}{m}\mathrm{Im}\left(C\bar{D}\right) = \frac{\hbar k}{m}\left(|A_2|^2 - |B_2|^2\right).$$

Away from the barrier (i.e., for x < 0 and x > a), the contribution $A_j e^{ikx}$ produces a right going flux, and $B_j e^{-ikx}$ a left going flux. The two remaining constants are determined by requiring a certain behaviour at infinity. We fix the incoming fluxes by choosing $A_1 = 1$, $B_2 = 0$. This should reflect a situation, where electrons move towards the barrier from the left. Then the part $B_j e^{-ikx}$ can be interpreted as a reflection, and the part $A_j e^{ikx}$ as a transmission. From the representations for the flux we obtain the equation,

$$R + T = 1$$

with the reflection coefficient R = $|B_1|^2$ and the transmission coefficient T = $|A_2|^2$. These can be interpreted as probabilities for the electron to be reflected and transmitted, respectively. The term tunnelling is used for the fact that the transmission coefficient is positive in general, i.e., the electron crosses the barrier with positive probability although its kinetic energy would lead to reflection in a classical description. Solving for the coefficients in the representation of the wave function gives,

$$B_1 = \frac{|z|^2 \left(e^{a\beta} - e^{-a\beta}\right)}{z^2 e^{-a\beta} - \bar{z}^2 e^{a\beta}},$$

showing that the reflection coefficient tends to zero with the width a of the barrier.

Wigner Transformation — The Classical Limit

In this, a connection between quantum and classical mechanics is established. We start with a nondimensionalization of the von Neumann equation,

$$ih\frac{\partial_Q}{\partial t} = -\frac{h^2}{2m}(\Delta xQ - \Delta yQ) + \left(W(x) - W(y)\right)Q.$$

Let W_0 denote a typical potential difference such that the dimensionless $W_s = W/W_0$ takes moderate values. Then we introduce reference values L_0 and t_0 for length and time, respectively, and the corresponding scaling $x_s = x/L_0$, $t_s = t/t_0$. Here L_0 is chosen as a length scale which is typical for an experiment. We assume that L_0 is large enough such that the action $L_0\sqrt{mW_0}$ is large compared to $\hbar$, i.e.,

$$\varepsilon = \frac{h}{L_0\sqrt{mW_0}} << 1.$$

Then t_0 is determined such that $\varepsilon = \hbar/(t_0 W_0)$. Carrying out the scaling in the von Neumann equation and dropping the subscripts s for simplicity, we obtain,

$$i\varepsilon\frac{\partial_Q}{\partial t} = -\frac{\varepsilon^2}{2}(\Delta xQ - \Delta yQ) + \left(W(x) - W(y)\right)Q.$$

We are interested in approximating this equation for small values of ε. This is made possible by the so called Wigner transformation (Eugene Wigner, 1932). It consists of two steps: First, a coordinate transformation in the 6-dimensional (x, y)-space is introduced:

$$x = \xi + \frac{\varepsilon\eta}{2}, y = \xi - \varepsilon\eta\ 2$$

In terms of (ξ, η), the von Neumann equation becomes,

$$i\frac{\partial_Q}{\partial t} = -\nabla_\eta \cdot \nabla_{\zeta Q} + \frac{W(\zeta + \varepsilon\eta/2) + \left(W(\zeta - \varepsilon\eta/2)\right)}{\varepsilon}Q.$$

The second step is a Fourier transformation: The Wigner function is introduced by,

$$W(\zeta, k, t) = (2\pi)^{-3} \int_{R3} Q(\zeta + \varepsilon\eta / 2, \zeta - \varepsilon\eta / 2, t) e^{-ik\cdot\eta} \, d\eta.$$

The inversion formula then gives,

$$Q(\zeta + \varepsilon\eta / 2, \zeta - \varepsilon\eta / 2, t) = \int_{R3} w(\zeta, k, t) e^{-ik\cdot\eta} \, dk =: \hat{w}(\zeta, \eta, t).$$

Observing $\nabla_\eta \hat{w} = ikw$, the Wigner function satisfies the Wigner equation

$$\frac{\partial w}{\partial t} + k \cdot \nabla_\xi w - (2\pi)^{-3} \int_{R^3} \frac{W(\xi + \varepsilon\eta / 2) - W(\xi - \varepsilon\eta / 2)}{i\varepsilon} \hat{w} e^{-ik\cdot\eta} \, d\eta = 0 .$$

In this equation, we may let the small parameter ε tend to zero. Observing $\eta\hat{w} = i\nabla_k w$, the classical limit $\varepsilon \to 0$ gives formally,

$$\frac{\partial wo}{\partial t} + k \cdot \nabla_\xi w_0 - \nabla_\xi W \cdot \nabla_k w_0 = 0 ,$$

for $w_0 = lim_{\varepsilon\to 0}\, w$

The physical significance of the Wigner function is easily seen for a pure state $Q(x, y, t) = \psi(x, t)\psi(y, t)$. Then the probability density for the position of the particle is given by,

$$|\psi(x, t)|^2 = 0(x, x, t) = \hat{w}(x, 0, t) = \int_{R^3} w(x, k, t) dk,$$

and the probability flux (or momentum density) by,

$$\varepsilon Im\left(\overline{\psi(x, t)} \nabla \psi(x, t)\right) = -i\nabla_\eta \hat{w}(x, 0, t) = \int_{R^3} kw(x, k, t) dk \ .$$

These two relations show that the new independent variable k can be interpreted as a momentum, and the Wigner function has some properties of a probability distribution in phase space (i.e., position-momentum space). Its zeroth order moment with respect to k is the probability density in position space and its first order moment the momentum density. However, there is one problem with this view. It can be shown that the Wigner function may take negative values in general.

In the classical limit, on the other hand, there is no problem like this. Solutions of $\frac{\partial wo}{\partial t} + k \cdot \nabla_\xi w_0 - \nabla_\xi W \cdot \nabla_k w_0 = 0$, remain positive if they are positive at some point in time. The Liouville equation $\frac{\partial wo}{\partial t} + k \cdot \nabla_\xi w_0 - \nabla_\xi W \cdot \nabla_k w_0 = 0$, is the correct equation for a probabilistic description of classical mechanics. Let (x(t), k(t)) denote a classical particle trajectory, i.e., a solution of,

$$\dot{x} = k \ , \dot{k} = -\nabla x W(x).$$

Then, for solutions $w_0(x, k, t)$ of $\frac{\partial wo}{\partial t} + k \cdot \nabla_\xi w_0 - \nabla_\xi W \cdot \nabla_k w_0 = 0$, $\frac{d}{dt} w_0(x(t), k(t), t) = 0$ holds.

Quantum Superposition

The superposition principle is the idea that a system is in all possible states at the same time, until it is measured. After measurement it then falls to one of the basis states that form the superposition, thus destroying the original configuration. The superposition principle explains the "quantum weirdness" observed with many experiments. A classic example of this is the double-slit experiment. Here, two slits in a barrier allow for the passage of (for example) electrons. The result of this experiment is an interference pattern not predicted by classical mechanics.

The most counter-intuitive aspect of quantum mechanics is the essential role that measurement plays in determining the behaviour of physical reality. To fully appreciate the counter-intuitive and paradoxical nature of quantum mechanics we study the two-slit experiment in some detail. The heart of quantum mechanics is tied down to the wave-particle duality of elementary particles. An elementary particle can be localized (captured) as if it were a point like particle; on the other hand it can exist everywhere just like a wave field which has an extended structure. In this section we shall illustrate the wave-attribute and particle-attribute of an elementary particle. In the macro-world, the concept of a particle is easy to comprehend. One starts with a piece of matter and keep on breaking it until one reaches the smallest constituent of matter. This smallest constituent is a particle. For example, the powder of a chalk can approximately be regarded as particles. In this way a particle is a point-like object. In geometry, a point has no size. However in physics one needs to measure 'size', that is, one needs a very powerful microscope in order to determine microscope size. The microscope in this scale is just the high energy accelerator and detector and at present energy the smallest size that we can measure is up to the order of 10^{-19} m. A wave is also easy to visualize in the macro-world, it is the motion of a disturbance and a simple wave is an extended entity with a periodic structure. Water wave, sound wave are just energy propagating in continuous media: water and air. Originally it was thought that light wave also needs the presence of a medium, the ether. This ether is ruled out by the constancy of the speed of light. In micro-world we cannot directly 'see' the particles of waves, so how do we extend the concept of particle and wave from macro-world to the micro-world? This is done by examining the behaviour of a physical system under interference experiments. In macro-world, bullets can be taken as particles. In the experiment as illustrated in figure below, a bullet from the firing gun can only go through either slit 1 or slit 2 and it is detected by the movable detector at the backstop. The experiment is first done by covering the slit 2 so that the bullet can go through only slit 1. After firing for a suitable time interval ΔT, the distribution of the bullets detected is plotted along the x-direction and the distribution curve $P_1(x)$ is obtained. The experiment is repeated with the same time interval ΔT with the slit 1 being covered instead of slit 2. The result is the distribution curve $P_2(x)$.

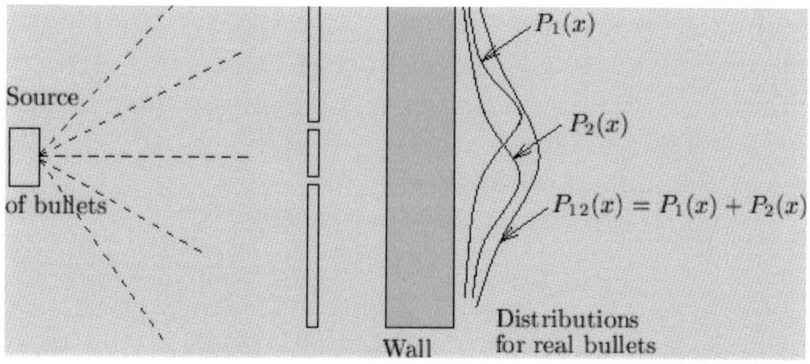

Interference Experiment with Bullets

When both slits 1 and 2 are open, the combined distribution curve $P_1(x)+P_2(x)$ is obtained, and it is found that is in fact a sum if $P_1(x)$ and $P_2(x)$. In this way we identify particle-like behavior for a physical system in the microwrold by the distribution curves $P_1(x)$, $P_2(x)$ and $P_1(x)$ $P_2(x)$. There is no interference in the sense that the probability that a bullet will reach a point x is by taking a path either through slit 1 or through slit 2, and this explains the final result being a sum of $P_1(x)$ and $P_2(x)$. A similar experiment is performed for water wave as shown infigure below. The detector can only measure the intensity I of the wave, which is proportional to the square of the height h of the wave. Let h_1 and h_2, respectively, be the amplitudes (heights) of the waves arriving at the detector when slit 2 and slit 1 are closed; we then have I_1 á $h_1^2(x)$ and I_2 á $h_2^2(x)$ are the intensity distributions of the water wave when slit 2 and slit 1 are closed respectively. When slits 1 and 2 are open, the resultant intensity is, from the superposition of waves, given by,

$$\begin{aligned} I_{12} &= (h_1 + h_2)^2 \\ &\neq I_1 + I_2 \end{aligned}$$

Recall the reason that the intensity of the interference pattern, namely $(h_1 + h_2)^2$, is not the sum of the individual amplitude is due to constructive and destructive interference, and which gives the characteristic minima and maxima of interference. Interference, as originally used by Young for light, is the best indication for a phenomenon being wave-like.

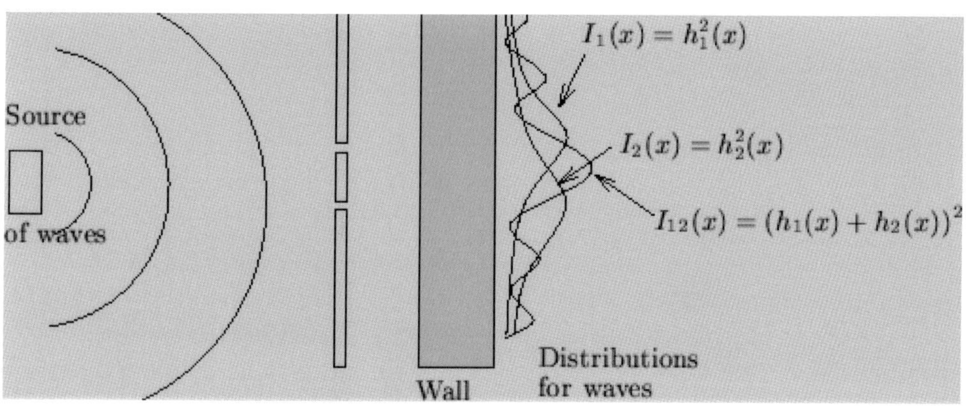

Waves

To ascertain whether an elementary particle such as an electron behaves like a wave or particle, we carry out the interference experiment similar to the one we have considered for water waves. What we need to state at the outset that the interference patterns $P_1 + P_2$ and I_{12} can both be obtained for the electron depending on how we perform the experiment. The experimental arrangement consists of an electron gun which sends identical electrons through a screen which has two slits to a wall where an apparatus keeps track of the point at which the electron stops. The electron gun produces the electrons one by one, so that at any given time there is only one electron traveling from the electron gun to the wall. We consider two different experiments with this arrangement, namely, one experiment in which a measurement is carried out to determine which slit the electron went through, and a second experiment in which no measurement is made to determine which slit the electron goes through. In both cases a large number, say 10^6, electrons are sent in, one by one, and the distribution of the positions at which the electrons hit the wall is measured.

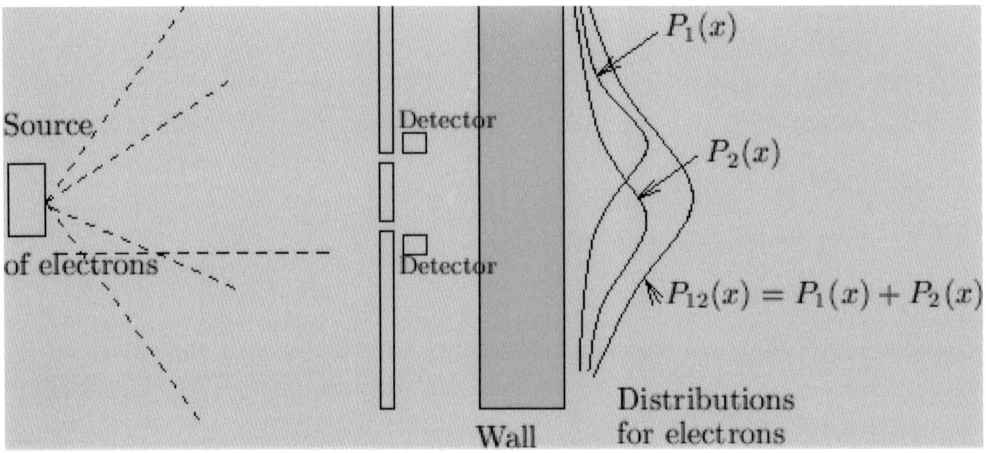

Electron with detectors

We perform the experiment as given in figure above, with both slits 1 and 2, open and with the additional requirement that we determine which slit the electron actually passes through. This can be arranged by fixing two detectors, say a light source, at the back of the slits as shown in figure above. Since we know which slit the electron goes through we can plot three distribution curves. P_1 and P_2 are the distribution curves for electrons go through slit 1 and slit 2 respectively. Similar to the result obtained for bullets, the probability of the electron arriving at a point on the wall when both slits are open, denoted by P_{12} is given by,

$$P_{12} = P_1 + P_2$$

P_{12} is the distribution curve for electrons that passes through either slit 1 or 2. We consequently have the result that when the electron's path is measured, it has a particle-like behavior. Experiment without Detection Consider now the same experiment as before, but with the detectors removed, as shown in figure below. In other words, we do not make any measurement to determine which slit the electron goes through. The result of this experiment is illustrated in figure below, and shows that a single electron gives rise to an interference. The interference pattern P_{12} is exactly like I_{12} as obtained for water waves. This suggests that electrons behave like waves and we have to introduce a probability amplitude ψ_1 for electrons when slit 2 is closed and an amplitude ψ_2 for electrons when slit 1 is closed. We then have in analogy with waves,

$$P_1 = |\psi_1|^2$$

$$P_2 = |\psi_2|^2$$

When both slits 1 and 2 are open, and no measurement is made, the resultant distribution P_{12} is the square of modulus of the sum of ψ_1 and ψ_2. The probability amplitudes obey the superposition principle when the different paths are not known, and yield,

$$\psi_{12} = \psi_1 + \psi_2$$

It is the superposed amplitude Ψ_{12} that determines the outcome when no measurement is performed. Hence,

$$P_{12} = |\psi_{12}|^2 = |\psi_1 + \psi_2|^2$$
$$\neq |\psi_1|^2 + |\psi_2|^2 = P_1 + P_2$$

The superposition principle is the unique feature of quantum mechanics, and shows graphically that, under some circumstances, particles behave as probability waves.

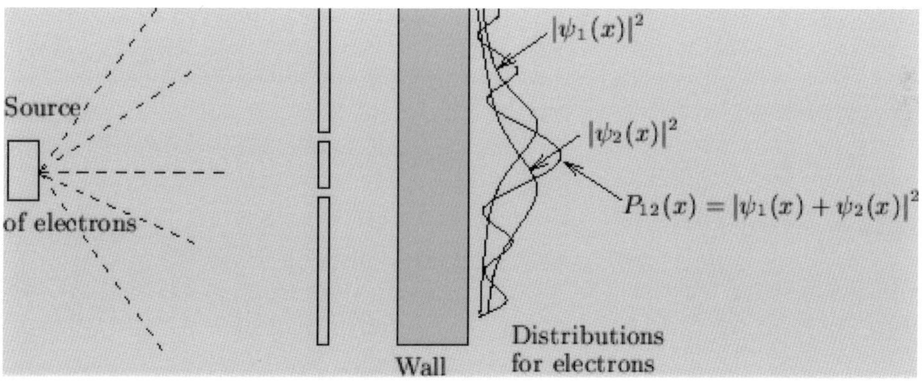

Electron without Detector

Note from figure above that the points of minima, say xM, of the interference pattern indicate that **no** electrons will be detected at those points. This is a remarkable, since if say only one slit was open there is a finite likelihood of an electron arriving at xM, but with both slits open, unlike the case for bullets, no electron can arrive there. This result is counter-intuitive since one would expect, as in the case of bullets, that for both slits open the electron would have two ways of arriving at point xM. In sum, when we do not observe which path the electron takes, it behaves like a wave. An actual interference experiment for atoms instead of electrons is given in figure below, and leads to the conclusion that the electron distribution curve can either be wave-like or particle-like depending on whether we require the information as to which slit the electron passes through.

Quantum Tunnelling

The quantum tunnelling effect is, as the name suggests, a quantum phenomenon which occurs when particles move through a barrier that should be impossible to move through according to classical physics. The barrier can be a physically impassable medium, like an insulator or a vacuum, or it can be a region of high potential energy.

In classical mechanics, if a particle has insufficient energy to overcome a potential barrier, it simply won't. In the quantum world, however, particles can often behave like waves. On encountering a barrier, a quantum wave will not end abruptly - its amplitude will decrease exponentially. This drop in amplitude corresponds to a drop in probability of finding a particle as you look further into the barrier. If the barrier is thin enough, the amplitude may be non-zero on the other side, so there is a finite probability that some of the particles will tunnel through the barrier.

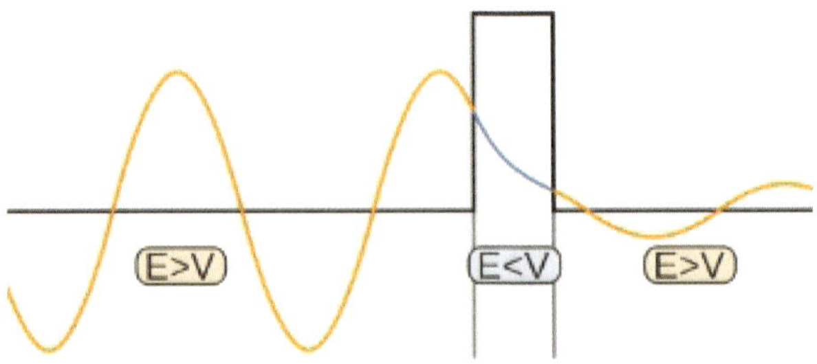

In regions where the potential energy is higher than the wave's energy, the amplitude of the wave decays exponentially. If the region is narrow enough, the wave can have a non-zero amplitude on the other side

The tunneling current is defined as the ratio of the current density emerging from the barrier divided by the current density incident on the barrier. If this transmission coefficient across the barrier is a non-zero value, there is a finite likelihood of a particle tunneling through the barrier.

Discovery of Quantum Tunneling

The possibility of tunneling was first noticed by F. Hund in 1927, whilst calculating the ground state energy in a "double-well" potential - a system where two separate states of similar energies are separated by a potential barrier. Many molecules, such as ammonia, are examples of this type of system. "Inversion" transitions between two geometric states are forbidden by classical mechanics, but are made possible by quantum tunneling.

In the same year, L. Nordheim noticed another incidence of the tunneling phenomenon whilst studying the reflection of electrons from a variety of surfaces. In the following few years, tunneling was successfully used to calculate the ionization rate of hydrogen by Oppenheimer, and to explain the range of alpha decay rates of radioactive nuclei by George Garnow, and, independently, R. W. Gurney and E. U. Condon.

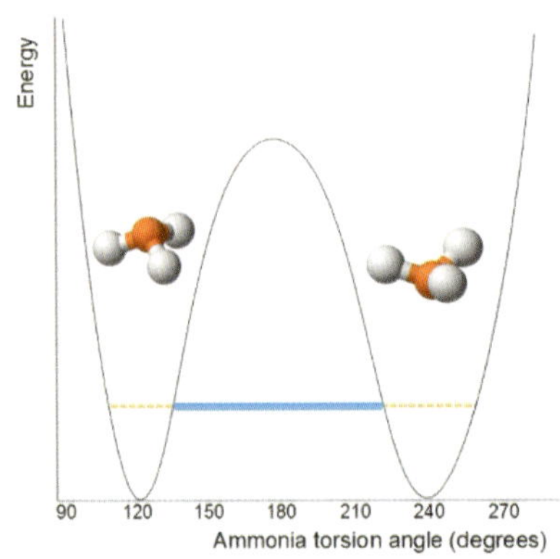

The two energetically equivalent states of ammonia, NH_3, can exchange more readily than is classically predicted - this is because the molecule can tunnel through the potential barrier at lower energies than are required to pass through the transition state

Quantum Tunneling in Nature

Although, like much of quantum physics, tunneling may appear to have little relevance to everyday life, it is a fundamental process of nature which is responsible for many things on which life itself is dependant.

It has even been hypothesized that the very beginning of the universe was caused by a tunneling event, allowing the universe to pass from a "state of no geometry" (no space or time) to a state in which space, time, matter, and life could exist.

Tunneling in Stellar Fusion

Fusion is the process by which small nuclei can join together to form larger nuclei, releasing huge amounts of energy. Fusion inside stars produces all the elements of the periodic table, except hydrogen, and fusion of hydrogen into helium is the process which gives stars their power.

However, fusion happens much more often than we originally thought it should. As all nuclei are positively charged, they repel each other very strongly, and their kinetic energy is very rarely sufficient to overcome this repulsion and allow fusion to occur. If tunneling effects are taken into account, however, the proportion of hydrogen nuclei which are able to undergo fusion increases dramatically.

This helps to explain how stars are able to remain stable for millions of years. The process is still not very likely though - an average hydrogen nucleus will undergo over 1000 head-on collisions before it fuses with another.

Tunneling in Smell Receptors

Until quite recently, it was believed that chemical receptors in the nose (400 different kinds in humans) detected the presence of various chemicals by a lock-and-key process, which identified the molecule's physical shape.

There are some issues with this theory, however. For example, ethanol and ethanethiol, which have very similar shapes, smell completely different (ethanol is the alcohol we drink - ethanethiol smells of rotten eggs). This suggests that some other identifying mechanism is at work.

A theory which has been growing in popularity over the last decade or so is that smell receptors rely in part on quantum tunneling to identify chemicals. The receptors pump a small current across the odourant molecule, causing it to vibrate in a characteristic way. In order for the current to flow, however, the electrons must tunnel through the non-conducting gap between the cells of the receptor and the molecule.

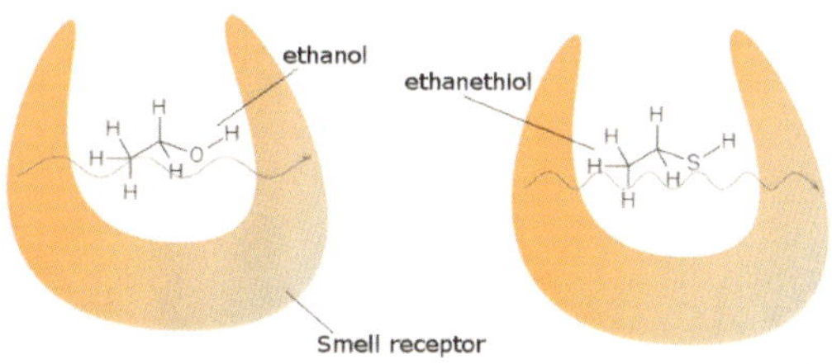

Smell receptors can detect the differences between similarly-shaped molecules by tunneling a small current across them, causing a characteristic vibration

Applications

Tunnelling occurs with barriers of thickness around 1-3 nm and smaller, but is the cause of some important macroscopic physical phenomena. For instance, tunnelling is a source of current leakage in very-large-scale integration (VLSI) electronics and results in the substantial power drain and heating effects that plague high-speed and mobile technology; it is considered the lower limit on how small computer chips can be made. Tunnelling is a fundamental technique used to program the floating gates of flash memory, which is one of the most significant inventions that have shaped consumer electronics in the last two decades.

Nuclear Fusion in Stars

Quantum tunnelling is essential for nuclear fusion in stars. Temperature and pressure in the core of stars are insufficient for nuclei to overcome the Coulomb barrier in order to achieve a thermonuclear fusion. However, there is some probability to penetrate the barrier due to quantum tunnelling. Though the probability is very low, the extreme number of nuclei in a star generates a steady fusion reaction over millions or even billions of years - a precondition for the evolution of life in insolation habitable zones.

Radioactive Decay

Radioactive decay is the process of emission of particles and energy from the unstable nucleus of an atom to form a stable product. This is done via the tunnelling of a particle out of the nucleus (an electron tunnelling into the nucleus is electron capture). This was the first application of quantum tunnelling and led to the first approximations. Radioactive decay is also a relevant issue for astro-biology as this consequence of quantum tunnelling is creating a constant source of energy over a large period of time for environments outside the circumstellar habitable zone where insolation would not be possible (subsurface oceans) or effective.

Astrochemistry in Interstellar Clouds

By including quantum tunnelling the astrochemical syntheses of various molecules in interstellar clouds can be explained such as the synthesis of molecular hydrogen, water (ice) and the prebiotic important formaldehyde.

Quantum Biology

Quantum tunnelling is among the central non trivial quantum effects in quantum biology. Here it is important both as electron tunnelling and proton tunnelling. Electron tunnelling is a key factor in many biochemical redox reactions (photosynthesis, cellular respiration) as well as enzymatic catalysis while proton tunnelling is a key factor in spontaneous mutation of DNA.

Spontaneous mutation of DNA occurs when normal DNA replication takes place after a particularly significant proton has defied the odds in quantum tunnelling in what is called "proton tunnelling" (quantum biology). A hydrogen bond joins normal base pairs of DNA. There exists a double well potential along a hydrogen bond separated by a potential energy barrier. It is believed that the double well potential is asymmetric with one well deeper than the other so the proton normally rests in the deeper well. For a mutation to occur, the proton must have tunnelled into the shallower of the two potential wells. The movement of the proton from its regular position is called a tautomeric transition. If DNA replication takes place in this state, the base pairing rule for DNA may be jeopardised causing a mutation. Per-Olov Lowdin was the first to develop this theory of spontaneous mutation within the double helix (quantum bio). Other instances of quantum tunnelling-induced mutations in biology are believed to be a cause of ageing and cancer.

Cold Emission

Cold emission of electrons is relevant to semiconductors and superconductor physics. It is similar to thermionic emission, where electrons randomly jump from the surface of a metal to follow a voltage bias because they statistically end up with more energy than the barrier, through random collisions with other particles. When the electric field is very large, the barrier becomes thin enough for electrons to tunnel out of the atomic state, leading to a current that varies approximately exponentially with the electric field. These materials are important for flash memory, vacuum tubes, as well as some electron microscopes.

Tunnel Junction

A simple barrier can be created by separating two conductors with a very thin insulator. These are tunnel junctions, the study of which requires quantum tunnelling. Josephson junctions take advantage of quantum tunnelling and the superconductivity of some semiconductors to create the Josephson effect. This has applications in precision measurements of voltages and magnetic fields, as well as the multijunction solar cell.

Quantum-dot Cellular Automata

QCA is a molecular binary logic synthesis technology that operates by the inter-island electron tunneling system. This is a very low power and fastest device that can operate at maximum 15 PHz of frequency.

Tunnel Diode

Diodes are electrical semiconductor devices that allow electric current flow in one direction more than the other. The device depends on a depletion layer between N-type and P-type semiconductors to serve its purpose; when these are very heavily doped the depletion layer can be thin enough for tunnelling. Then, when a small forward bias is applied the current due to tunnelling is significant. This has a maximum at the point where the voltage bias is such that the energy level of the p and n conduction bands are the same. As the voltage bias is increased, the two conduction bands no longer line up and the diode acts typically.

Because the tunnelling current drops off rapidly, tunnel diodes can be created that have a range of voltages for which current decreases as voltage is increased. This peculiar property is used in some applications, like high speed devices where the characteristic tunnelling probability changes as rapidly as the bias voltage.

The resonant tunnelling diode makes use of quantum tunnelling in a very different manner to achieve a similar result. This diode has a resonant voltage for which there is a lot of current that favors a particular voltage, achieved by placing two very thin layers with a high energy conductance band very near each other. This creates a quantum potential well that has a discrete lowest energy level. When this energy level is higher than that of the electrons, no tunnelling will occur, and the diode is in reverse bias. Once the two voltage energies align, the electrons flow like an open wire. As the voltage is increased further tunnelling becomes improbable and the diode acts like a normal diode again before a second energy level becomes noticeable.

Tunnel Field-effect Transistors

A European research project has demonstrated field effect transistors in which the gate (channel) is controlled via quantum tunnelling rather than by thermal injection, reducing gate voltage from ~1 volt to 0.2 volts and reducing power consumption by up to 100×. If these transistors can be scaled up into VLSI chips, they will significantly improve the performance per power of integrated circuits.

Quantum Conductivity

While the Drude model of electrical conductivity makes excellent predictions about the nature of electrons conducting in metals, it can be furthered by using quantum tunnelling to explain the nature of the electron's collisions. When a free electron wave packet encounters a long array of uniformly spaced barriers the reflected part of the wave packet interferes uniformly with the transmitted one between all barriers so that there are cases of 100% transmission. The theory predicts that if positively charged nuclei form a perfectly rectangular array, electrons will tunnel through the metal as free electrons, leading to an extremely high conductance, and that impurities in the metal will disrupt it significantly.

Scanning Tunnelling Microscope

The scanning tunnelling microscope (STM), invented by Gerd Binnig and Heinrich Rohrer, may allow imaging of individual atoms on the surface of a material. It operates by taking advantage of the relationship between quantum tunnelling with distance. When the tip of the STM's needle is brought very close to a conduction surface that has a voltage bias, by measuring the current of electrons that are tunnelling between the needle and the surface, the distance between the needle and the surface can be measured. By using piezoelectric rods that change in size when voltage is applied over them the height of the tip can be adjusted to keep the tunnelling current constant. The time-varying voltages that are applied to these rods can be recorded and used to image the surface of the conductor. STMs are accurate to 0.001 nm, or about 1% of atomic diameter.

Faster than Light

Some physicists have claimed that it is possible for spin-zero particles to travel faster than the speed of light when tunnelling. This apparently violates the principle of causality, since there will be a frame of reference in which it arrives before it has left. In 1998, Francis E. Low reviewed briefly the phenomenon of zero-time tunnelling. More recently experimental tunnelling time data of phonons, photons, and electrons have been published by Günter Nimtz.

Other physicists, such as Herbert Winful, have disputed these claims. Winful argues that the wavepacket of a tunnelling particle propagates locally, so a particle can't tunnel through the barrier non-locally. Winful also argues that the experiments that are purported to show non-local propagation have been misinterpreted. In particular, the group velocity of a wavepacket does not measure its speed, but is related to the amount of time the wavepacket is stored in the barrier.

Mathematical Discussions of Quantum Tunnelling

The following subsections discuss the mathematical formulations of quantum tunnelling.

The Schrödinger Equation

The time-independent Schrödinger equation for one particle in one dimension can be written as,

$$-\frac{\hbar^2}{2m}\frac{d^2}{dx^2}\Psi(x)+V(x)\Psi(x)=E\Psi(x) \text{ or}$$

$$\frac{d^2}{dx^2}\Psi(x)=\frac{2m}{\hbar^2}(V(x)-E)\Psi(x)\equiv\frac{2m}{\hbar^2}M(x)\Psi(x),$$

where $\hbar$ is the reduced Planck's constant, m is the particle mass, x represents distance measured in the direction of motion of the particle, Ψ is the Schrödinger wave function, V is the potential energy of the particle (measured relative to any convenient reference level), E is the energy of the particle that is associated with motion in the x-axis (measured relative to V), and M(x) is a quantity defined by V(x) – E which has no accepted name in physics.

The solutions of the Schrödinger equation take different forms for different values of x, depending on whether M(x) is positive or negative. When M(x) is constant and negative, then the Schrödinger equation can be written in the form,

$$\frac{d^2}{dx^2}\Psi(x)=\frac{2m}{\hbar^2}M(x)\Psi(x)=-k^2\Psi(x), \quad \text{where } k^2=-\frac{2m}{\hbar^2}M.$$

The solutions of this equation represent travelling waves, with phase-constant $+k$ or $-k$. Alternatively, if M(x) is constant and positive, then the Schrödinger equation can be written in the form,

$$\frac{d^2}{dx^2}\Psi(x)=\frac{2m}{\hbar^2}M(x)\Psi(x)=\kappa^2\Psi(x), \quad \text{where } \kappa^2=\frac{2m}{\hbar^2}M.$$

The solutions of this equation are rising and falling exponentials in the form of evanescent waves. When M(x) varies with position, the same difference in behaviour occurs, depending on whether M(x) is negative or positive. It follows that the sign of M(x) determines the nature of the medium, with negative M(x) corresponding to medium A as described above and positive M(x) corresponding to medium B. It thus follows that evanescent wave coupling can occur if a region of positive M(x) is sandwiched between two regions of negative M(x), hence creating a potential barrier.

The mathematics of dealing with the situation where M(x) varies with x is difficult, except in special cases that usually do not correspond to physical reality. A discussion of the semi-classical approximate method, as found in physics textbooks. A full and complicated mathematical treatment appears in the 1965 monograph by Fröman and Fröman noted below. Their ideas have not been incorporated into physics textbooks, but their corrections have little quantitative effect.

The WKB Approximation

The wave function is expressed as the exponential of a function:

$\Psi(x)=e^{\Phi(x)}$, where $\Phi''(x)+\Phi'(x)^2=\frac{2m}{\hbar^2}(V(x)-E)$.

$\Phi'(x)$ is then separated into real and imaginary parts:

$\Phi'(x)=A(x)+iB(x)$, , where $A(x)$ and $B(x)$ are real-valued functions.

Substituting the second equation into the first and using the fact that the imaginary part needs to be 0 results in:

$$A'(x)+A(x)^2-B(x)^2=\frac{2m}{\hbar^2}(V(x)-E).$$

To solve this equation using the semiclassical approximation, each function must be expanded as a power series in $\hbar$. From the equations, the power series must start with at least an order of $\hbar^{-1}$ to satisfy the real part of the equation; for a good classical limit starting with the highest power of Planck's constant possible is preferable, which leads to,

$$A(x)=\frac{1}{\hbar}\sum_{k=0}^{\infty}\hbar^k A_k(x)$$

and

$$B(x)=\frac{1}{\hbar}\sum_{k=0}^{\infty}\hbar^k B_k(x),$$

with the following constraints on the lowest order terms,

$$A_0(x)^2-B_0(x)^2=2m(V(x)-E)$$

and

$$A_0(x)B_0(x)=0.$$

At this point two extreme cases can be considered.

Case 1

If the amplitude varies slowly as compared to the phase $A_0(x)=0$ and

$$B_0(x)=\pm\sqrt{2m(E-V(x))}$$

which corresponds to classical motion. Resolving the next order of expansion yields,

$$\Psi(x) \approx C \frac{e^{i\int dx \sqrt{\frac{2m}{\hbar^2}(E-V(x))}+\theta}}{\sqrt[4]{\frac{2m}{\hbar^2}(E-V(x))}}$$

Case 2

If the phase varies slowly as compared to the amplitude, $B_0(x)=0$, and

$$A_0(x)=\pm\sqrt{2m(V(x)-E)}$$

which corresponds to tunnelling. Resolving the next order of the expansion yields,

$$\Psi(x) \approx \frac{C_+ e^{+\int dx \sqrt{\frac{2m}{\hbar^2}(V(x)-E)}} + C_- e^{-\int dx \sqrt{\frac{2m}{\hbar^2}(V(x)-E)}}}{\sqrt[4]{\frac{2m}{\hbar^2}(V(x)-E)}}$$

In both cases it is apparent from the denominator that both these approximate solutions are bad near the classical turning points $E=V(x)$. Away from the potential hill, the particle acts similar to a free and oscillating wave; beneath the potential hill, the particle undergoes exponential changes in amplitude. By considering the behaviour at these limits and classical turning points a global solution can be made.

To start, choose a classical turning point, x_1 and expand $\frac{2m}{\hbar^2}(V(x)-E)$ in a power series about x_1:

$$\frac{2m}{\hbar^2}(V(x)-E)=v_1(x-x_1)+v_2(x-x_1)^2+\cdots$$

Keeping only the first order term ensures linearity:

$$\frac{2m}{\hbar^2}(V(x)-E)=v_1(x-x_1).$$

Using this approximation, the equation near x_1 becomes a differential equation:

$$\frac{d^2}{dx^2}\Psi(x)=v_1(x-x_1)\Psi(x).$$

This can be solved using Airy functions as solutions.

$$\Psi(x)=C_A Ai\left(\sqrt[3]{v_1}(x-x_1)\right)+C_B Bi\left(\sqrt[3]{v_1}(x-x_1)\right)$$

Taking these solutions for all classical turning points, a global solution can be formed that links the limiting solutions. Given the two coefficients on one side of a classical turning point, the two coefficients on the other side of a classical turning point can be determined by using this local solution to connect them.

Hence, the Airy function solutions will asymptote into sine, cosine and exponential functions in the proper limits. The relationships between C,θ and C_+,C_- are,

$$C_+ = \frac{1}{2} C \cos\left(\theta - \frac{\pi}{4}\right)$$

and

$$C_- = -C \sin\left(\theta - \frac{\pi}{4}\right)$$

With the coefficients found, the global solution can be found. Therefore, the transmission coefficient for a particle tunnelling through a single potential barrier is,

$$T(E) = e^{-2\int_{x_1}^{x_2} dx \sqrt{\frac{2m}{\hbar^2}[V(x)-E]}},$$

where x_1,x_2 are the two classical turning points for the potential barrier.

For a rectangular barrier, this expression is simplified to:

$$T(E) = e^{-2\sqrt{\frac{2m}{\hbar^2}(V_0-E)(x_2-x_1)}} = \tilde{V}_0^{-(x_2-x_1)}.$$

Uncertainty Principle

The uncertainty principle also called the Heisenberg Uncertainty Principle, or Indeterminacy Principle, articulated (1927) by the German physicist Werner Heisenberg, that the position and the velocity of an object cannot both be measured exactly, at the same time, even in theory. The very concepts of exact position and exact velocity together, in fact, have no meaning in nature.

Ordinary experience provides no clue of this principle. It is easy to measure both the position and the velocity of, say, an automobile, because the uncertainties implied by this principle for ordinary objects are too small to be observed. The complete rule stipulates that the product of the uncertainties in position and velocity is equal to or greater than a tiny physical quantity, or constant (about 10^{-34} joule-second, the value of the quantity h (where h is Planck's constant). Only for the exceedingly small masses of atoms and subatomic particles does the product of the uncertainties become significant.

Any attempt to measure precisely the velocity of a subatomic particle, such as an electron, will knock it about in an unpredictable way, so that a simultaneous measurement of its position has no validity. This result has nothing to do with inadequacies in the measuring instruments, the technique, or the observer; it arises out of the intimate connection in nature between particles and waves in the realm of subatomic dimensions.

Every particle has a wave associated with it; each particle actually exhibits wavelike behavior. The particle is most likely to be found in those places where the undulations of the wave are greatest, or most intense. The more intense the undulations of the associated wave become, however, the more ill defined becomes the wavelength, which in turn determines the momentum of the particle. So a strictly localized wave has an indeterminate wavelength; its associated particle, while having a definite position, has no certain velocity. A particle wave having a well-defined wavelength, on the other hand, is spread out; the associated particle, while having a rather precise velocity, may be almost anywhere. A quite accurate measurement of one observable involves a relatively large uncertainty in the measurement of the other.

The uncertainty principle is alternatively expressed in terms of a particle's momentum and position. The momentum of a particle is equal to the product of its mass times its velocity. Thus, the product of the uncertainties in the momentum and the position of a particle equals h/(2) or more. The principle applies to other related (conjugate) pairs of observables, such as energy and time: the product of the uncertainty in an energy measurement and the uncertainty in the time interval during which the measurement is made also equals h/(2) or more. The same relation holds, for an unstable atom or nucleus, between the uncertainty in the quantity of energy radiated and the uncertainty in the lifetime of the unstable system as it makes a transition to a more stable state.

The uncertainty principle, developed by W. Heisenberg, is a statement of the effects of wave-particle duality on the properties of subatomic objects. Consider the concept of momentum in the wave-like microscopic world. The momentum of wave is given by its wavelength. A wave packet like a photon or electron is a composite of many waves. Therefore, it must be made of many momentums. But how can an object have many momentums?

Of course, once a measurement of the particle is made, a single momentum is observed. But, like fuzzy position, momentum before the observation is intrinsically uncertain. This is what is know as the uncertainty principle, that certain quantities, such as position, energy and time, are unknown, except by probabilities. In its purest form, the uncertainty principle states that accurate knowledge of complementarity pairs is impossible. For example, you can measure the location of an electron, but not its momentum (energy) at the same time.

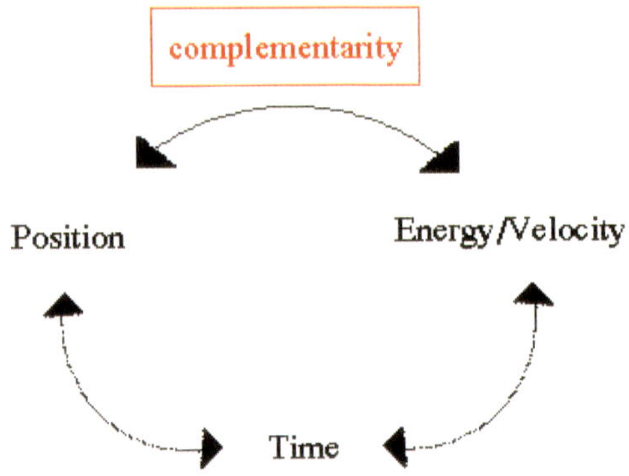

A characteristic feature of quantum physics is the principle of complementarity, which "implies the impossibility of any sharp separation between the behavior of atomic objects and the interaction with the measuring instruments which serve to define the conditions under which the phenomena appear." As a result, "evidence obtained under different experimental conditions cannot be comprehended within a single picture, but must be regarded as complementary in the sense that only the totality of the phenomena exhausts the possible information about the objects." This interpretation of the meaning of quantum physics, which implied an altered view of the meaning of physical explanation, gradually came to be accepted by the majority of physicists during the 1930's.

Mathematically we describe the uncertainty principle as the following, where 'x' is position and 'p' is momentum:

$$\Delta x \Delta p > \frac{h}{2\pi}$$

This is perhaps the most famous equation next to $E=mc^2$ in physics. It basically says that the combination of the error in position times the error in momentum must always be greater than Planck's constant. So, you can measure the position of an electron to some accuracy, but then its momentum will be inside a very large range of values. Likewise, you can measure the momentum precisely, but then its position is unknown.

Notice that this is not the measurement problem in another form, the combination of position, energy (momentum) and time are actually undefined for a quantum particle until a measurement is made (then the wave function collapses).

> Also notice that the uncertainty principle is unimportant to macroscopic objects since Planck's constant, h, is so small (10^{-34}). For example, the uncertainty in position of a thrown baseball is 10^{-30} millimeters.
>
> The depth of the uncertainty principle is realized when we ask the question; is our knowledge of reality unlimited? The answer is no, because the uncertainty principle states that there is a built-in uncertainty, indeterminacy, unpredictability to nature.

Quantum Entanglement

Quantum entanglement is a quantum mechanical phenomenon in which the quantum states of two or more objects have to be described with reference to each other, even though the individual objects may be spatially separated.

This leads to correlations between observable physical properties of the systems.

For example, it is possible to prepare two particles in a single quantum state such that when one is observed to be spin-up, the other one will always be observed to be spin-down and vice versa, this despite the fact that it is impossible to predict, according to quantum mechanics, which set of measurements will be observed.

As a result, measurements performed on one system seem to be instantaneously influencing other systems entangled with it.

But quantum entanglement does not enable the transmission of classical information faster than the speed of light.

Quantum entanglement has applications in the emerging technologies of quantum computing and quantum cryptography, and has been used to realize quantum teleportation experimentally.

At the same time, it prompts some of the more philosophically oriented discussions concerning quantum theory.

The correlations predicted by quantum mechanics, and observed in experiment, reject the principle of local realism, which is that information about the state of a system should only be mediated by interactions in its immediate surroundings.

Different views of what is actually occurring in the process of quantum entanglement can be related to different interpretations of quantum mechanics.

EPR Paradox and Stern-Gerlach Experiments

The original Stern-Gerlach experiment examined the behavior of silver atoms in magnetic fields. Magnetic dipoles couple to magnetic fields; they feel a force $F = \nabla\left(\vec{\mu}\cdot\vec{B}\right)$ where μ is the magnetic dipole moment and B is the magnetic field vector. Since silver atoms only have one valence electron with zero orbital angular momentum (a 5s electron), any coupling of silver atoms to magnetic fields is due to an intrinsic magnetic dipole moment or spin of the electron. Since only the behavior of the valence electron matters, the rest will refer to electrons where silver atoms were actually used originally.

In the Stern-Gerlach experiment, the magnetic field gradient ∇B is aligned along the z-axis and set to some constant C so that electrons are deflected in the magnetic field according to the magnetic dipole moment: $F = C\mu$. If the electrons are spin-up in the z-direction, they are deflected upwards, whereas if the electrons are spin-down in the z-direction, they are deflected downwards. This experiment provided the first evidence for the quantization of spin: on a detecting screen behind the magnets, only two peaks appear rather than a continuous spectrum, corresponding to the two possible quantized values of spin rather than a continuous distribution of possible magnetic dipole moments.

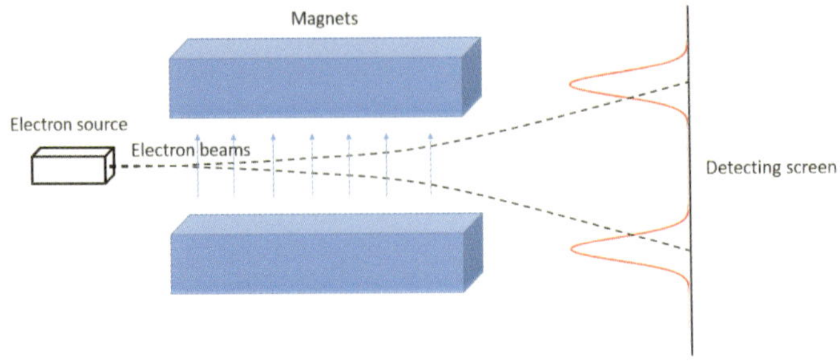

Experimental setup of a Stern-Gerlach experiment. An electron source beams electrons into a magnetic gradient, where they are deflected according to their intrinsic magnetic moment. Only two peaks are detected beyond the magnets, corresponding to the two possible spin values

The entirety of a Stern-Gerlach experiment can be thought of as a black box which is able to measure whether an entering electron is spin-up or spin down.

In the EPR thought experiment, Einstein, Podolsky, and Rosen considered a system of two electrons which are sent in opposite directions through two Stern-Gerlach apparatuses, measuring each of their spins. If the electrons are entangled, then when electron 1 is measured to be spin-up, electron 2 is measured to be spin-down a large percentage of the time, and vice versa. The precise meaning of "entangled" in quantum mechanics is discussed below mathematically. The result seemed to imply that at one Stern-Gerlach apparatus, the measurement result was being communicated instantly to the other apparatus. Although EPR did not perform the experiment themselves, actual experimental results are consistent with their mathematics. The nonlocality of the EPR paradox would be in violation of the principles of relativity, since it seems like information is being communicated to a distant location faster than the speed of light.

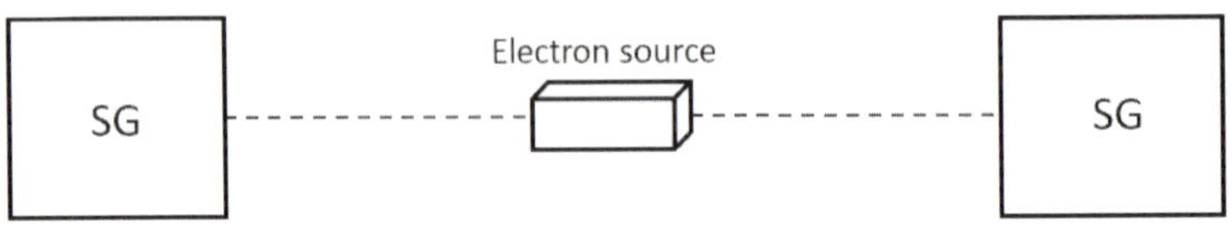

Setup of EPR's thought experiments: spin-entangled electrons are sent in opposite directions to two distant Stern-Gerlach apparatuses, where their spins are measured, revealing remarkable correlations

EPR tried to avoid this conclusion by asserting that there must have been some local hidden variables: at the electron source, some variable that cannot be measured was determining the outcomes of spin measurements when the entangled electrons were produced. They took this to mean that the theory of quantum mechanics was not complete, since the spin and spatial wavefunctions of the electrons did not include the hidden variables and so could not account for all the physics that they would observe. However, in the 1960s, Bell's theorem showed that this was not the case: no local hidden variables theory could possibly account for the degree of correlation present in measurement of spin-entangled electrons.

Entanglement usually originates on the subatomic scale via processes that produce two electrons simultaneously in a correlated way, although complicated systems of fiber optics can be used to produce e.g. systems of two photons with entangled polarizations.

Measurement of Electron Spin

A single electron spin in quantum mechanics is described by a vector in a two-dimensional vector space. The basis vectors are taken to be the eigenvectors of the operator giving the spin of the electron along the *z* axis, and are labeled $|0\rangle$ and $|1\rangle$, $|+\rangle$ and $|-\rangle$, or $|\uparrow\rangle$ and $|\downarrow\rangle$ depending on context. In this, the latter notation will be used. In general, an arbitrary electron spin state can be a superposition of the "spin-up" and "spin-down" states:

$$|\psi\rangle = c_1|\uparrow\rangle + c_2|\downarrow\rangle,$$

where c_1 and c_2 are some complex constants normalized so that $|c_1|^2 + |c_2|^2 = 1$. This notation suggests that a single-spin state can be written as a vector:

$$|\psi\rangle = \begin{pmatrix} c_1 \\ c_2 \end{pmatrix}.$$

The spin states are defined so that the spin operator in the z-direction,

$$\hat{S}_z = \frac{\hbar}{2}\begin{pmatrix} 1 & 0 \\ 0 & -1 \end{pmatrix},$$

$$\hat{S}_z \otimes I$$

has eigenvalues $\pm\frac{\hbar}{2}$ corresponding to the eigenvectors $|\uparrow\rangle$ and $|\downarrow\rangle$. This operator acts on the vector representation of spin states by simple matrix multiplication.

Entanglement and Tensor Products

In a two-particle system, the spins of two electrons can be described jointly using tensor product spaces. The tensor product of two electron spin states lives in a *four-dimensional* vector space, with basis vectors labeled:

$$|\uparrow\otimes|\uparrow\rangle, |\uparrow\rangle\otimes|\downarrow\rangle, |\downarrow\otimes|\uparrow\rangle, |\downarrow\otimes|\downarrow\rangle.$$

The tensor product of two states $|x\rangle$ and $|y\rangle$ obeys the following rules:

1) Scaling the tensor product is equivalent to scaling either state:

$$c\left(|x\rangle\otimes|y\rangle\right) = \left(c|x\rangle\right)\otimes|y\rangle = |x\rangle\otimes c|y\rangle.$$

2) The tensor product is distributive in both the first and second slots:

$$|x\rangle\otimes\left(|y_1\rangle + |y_2\rangle\right) = |x\rangle\otimes|y_1\rangle + |x\rangle\otimes|y_2\rangle$$

$$\left(|x_1\rangle + |x_2\rangle\right)\otimes|y\rangle = |x_1\rangle\otimes|y\rangle + |x_2\rangle\otimes|y\rangle$$

An arbitrary tensor product state for a two-spin system can be written out in the four-dimensional basis as:

$$|\psi\rangle c_{11}|\uparrow\rangle\otimes|\uparrow\rangle+c_{12}|\uparrow\rangle\otimes|\downarrow\rangle+c_{21}|\downarrow\rangle\otimes|\uparrow\rangle+c_{22}|\downarrow\rangle\otimes|\downarrow\rangle.$$

This notation suggests that it is more convenient to represent a two-spin state as a matrix, in analogue to the vector representation of a one-spin state:

$$|\psi\rangle=\begin{pmatrix} c_{11} & c_{12} \\ c_{21} & c_{22} \end{pmatrix}.$$

To perform a measurement on a two-spin system, one needs to specify the operators acting on each independent space. For instance, to measure the spin of just the first particle, use the operator written as $\hat{S}_z\otimes I$, which measures the spin of the first particle in the z-direction and acts on the second space as the identity. To measure the spins in the z-direction of both particles simultaneously, use $\hat{S}_z\otimes\hat{S}_z$ and so on.

Having defined this notation, it is now possible to state an explicit mathematical criterion of entanglement that is useful for proving things in quantum information theory.

> A two-electron spin state is a product state if it can be written as the tensor product of exactly two independent single-spin states.

Some states are obviously product states: $|\uparrow\rangle\otimes|\downarrow\rangle$, for instance, is clearly the product of $|\uparrow\rangle$ and $|\downarrow\rangle$. This state, therefore, can be described just by taking each independent single-spin state independently and ignoring the other. A state such as $|\uparrow\rangle\otimes|\downarrow\rangle+|\downarrow\rangle\otimes|\uparrow\rangle$, however, is clearly not a product state. Not all cases are so obvious, however, which is why formulating a precise mathematical criterion is important.

The intuition above suggests the definition of entanglement:

> A two-electron spin state is entangled if it is not a product state, that is, if the matrix representation M of the state does not satisfy $\det M=0$,

For states of more than two electrons, this condition is actually too strong -- if *det* $M\neq 0$, then *all* of the electron spins are entangled. But it's enough for just two of the electrons to be entangled in a multiparticle state for the whole state to be considered entangled. The more accurate condition for a general multiparticle state is that rank $M\geq 2$, since this is equivalent to entanglement in a two-dimensional subspace. Note that this condition reduces to *det* $M\neq 0$ for a two-electron spin state.

If a two-spin state can be written as a product state, it is as if each spin is independent. Measurements on each spin do not affect the other. But if a two-spin state is entangled, the fact that the the state is not a product state means that collapsing to one eigenvector in one space causes collapse in the other space as well. Measurements on one spin thus affect any entangled spins instantaneously by influencing what measurements of the other spin are possible, regardless of the spatial location of the other electron! It is important to note, however, that this is not in violation of special relativity: information *cannot* be sent faster than lightspeed via entanglement; the effects of entanglement can only be seen after the fact when two observers meet up to compare spin measurements.

Example

Consider the entangled state given by:

$$|\psi_E\rangle = \frac{1}{\sqrt{2}}\left(|\uparrow\rangle \otimes |\downarrow\rangle - |\downarrow\rangle \otimes |\uparrow\rangle\right),$$

and the non-entangled state given by,

$$|\psi_N\rangle = \frac{1}{\sqrt{2}}\left(|\uparrow\rangle \otimes |\downarrow\rangle + |\downarrow\rangle \otimes |\uparrow\rangle\right),$$

Show that measuring the spin of the first particle for $|\psi_E\rangle$ affects the probabilities of finding the second particle in each possible state, while the same measurement for $|\psi_N\rangle$ does not affect the results.

Solution:

For the entangled state, before measurement the second particle has an equal chance to be found in either the spin-up or spin-down states. After measurement of the first particle's spin with the operator $\hat{S}_z \otimes I$, if the first electron is found in the spin-up state, $|\psi_E\rangle$ collapses to $|\uparrow\rangle \otimes |\downarrow\rangle$ and the second electron *must* be found in the spin-down state. If the first electron is found in the spin-down state the reverse is true; thus, measurement of the state of the first electron has affected the overall state of the system.

For the non-entangled state, the probability of finding the first particle in the spin-up state is one. After measurement, the state does not change, and the second particle is still in an equal superposition of both spin states. Note that there also exist non-entangled states for which measuring the first particle would change the overall spin wavefunction without changing the state of the second particle.

Concept

Entanglement

An entangled system is defined to be one whose quantum state cannot be factored as a product of states of its local constituents; that is to say, they are not individual particles but are an inseparable whole. In entanglement, one constituent cannot be fully described without considering the other(s). Note that the state of a composite system is always expressible as a *sum*, or superposition, of products of states of local constituents; it is entangled if this sum necessarily has more than one term.

Quantum systems can become entangled through various types of interactions. Entanglement is broken when the entangled particles decohere through interaction with the environment; for example, when a measurement is made.

As an example of entanglement: a subatomic particle decays into an entangled pair of other particles. The decay events obey the various conservation laws, and as a result, the measurement outcomes of one daughter particle must be highly correlated with the measurement outcomes of the other daughter particle (so that the total momenta, angular momenta, energy, and so forth remains roughly the same before and after this process). For instance, a spin-zero particle could decay into a pair of spin-½ particles. Since the total spin before and after this decay must be zero (conservation of angular momentum), whenever the first particle is measured to be spin up on some axis, the other, when measured on the same axis, is always found to be spin down. (This is called the *spin anti-correlated* case; and if the prior probabilities for measuring each spin are equal, the pair is said to be in the singlet state.)

The special property of entanglement can be better observed if we separate the said two particles. Let's put one of them in the White House in Washington and the other in Buckingham Palace (think about this as a thought experiment, not an actual one). Now, if we measure a particular characteristic of one of these particles (say, for example, spin), get a result, and then measure the other particle using the same criterion (spin along the same axis), we find that the result of the measurement of the second particle will match (in a complementary sense) the result of the measurement of the first particle, in that they will be opposite in their values.

The above result may or may not be perceived as surprising. A classical system would display the same property, and a hidden variable theory would certainly be *required* to do so, based on conservation of angular momentum in classical and quantum mechanics alike. The difference is that a classical system has definite values for all the observables all along, while the quantum system does not. In a sense to be discussed below, the quantum system considered here seems to *acquire* a probability distribution for the outcome of a measurement of the spin along *any* axis of the *other* particle upon measurement of the *first* particle. This probability distribution is in general *different* from what it would be *without* measurement of the first particle. This may certainly be perceived as surprising in the case of spatially separated entangled particles.

Paradox

The paradox is that a measurement made on either of the particles apparently collapses the state of the entire entangled system—and does so instantaneously, before any information about the measurement result could have been communicated to the other particle (assuming that information cannot travel faster than light) and hence assured the "proper" outcome of the measurement of the other part of the entangled pair. In the Copenhagen interpretation, the result of a spin measurement on one of the particles is a collapse into a state in which each particle has a definite spin (either up or down) along the axis of measurement. The outcome is taken to be random, with each possibility having a probability of 50%. However, if both spins are measured along the same axis, they are found to be anti-correlated. This means that the random outcome of the measurement made on one particle seems to have been transmitted to the other, so that it can make the "right choice" when it too is measured.

The distance and timing of the measurements can be chosen so as to make the interval between the two measurements spacelike, hence, any causal effect connecting the events would have to travel faster than light. According to the principles of special relativity, it is not possible for any information to travel between two such measuring events. It is not even possible to say which of the measurements came first. For two spacelike separated events x_1 and x_2 there are inertial frames in which x_1 is first and others in which x_2 is first. Therefore, the correlation between the two measurements cannot be explained as one measurement determining the other: different observers would disagree about the role of cause and effect.

Hidden Variables Theory

A possible resolution to the paradox is to assume that quantum theory is incomplete, and the result of measurements depends on predetermined "hidden variables". The state of the particles being measured contains some hidden variables, whose values effectively determine, right from the moment of separation, what the outcomes of the spin measurements are going to be. This would mean that each particle carries all the required information with it, and nothing needs to be transmitted from one particle to the other at the time of measurement. Einstein and others must be incomplete. (In fact similar paradoxes can arise even without entanglement: the position of a single particle is spread out over space, and two widely separated detectors attempting to detect the particle in two different places must instantaneously attain appropriate correlation, so that they do not *both* detect the particle.)

Violations of Bell's Inequality

The hidden variables theory fails, however, when we consider measurements of the spin of entangled particles along different axes (for example, along any of three axes that make angles of 120 degrees). If a large number of pairs of such measurements are made (on a large number of pairs of entangled particles), then statistically, if the local realist or hidden variables view were correct, the results would always satisfy Bell's inequality. A number of experiments have shown in practice that Bell's inequality is not satisfied. However, prior to 2015, all of these had loophole problems that were considered the most important by the community of physicists. When measurements of the entangled particles are made in moving relativistic reference frames, in which each measurement (in its own relativistic time frame) occurs before the other, the measurement results remain correlated.

The fundamental issue about measuring spin along different axes is that these measurements cannot have definite values at the same time—they are incompatible in the sense that these measurements' maximum simultaneous precision is constrained by the uncertainty principle. This is contrary to what is found in classical physics, where any number of properties can be measured simultaneously with arbitrary accuracy. It has been proven mathematically that compatible measurements cannot show Bell-inequality-violating correlations, and thus entanglement is a fundamentally non-classical phenomenon.

Other Types of Experiments

In experiments in 2012 and 2013, polarization correlation was created between photons that never coexisted in time. The authors claimed that this result was achieved by entanglement swapping between two pairs of entangled photons *after* measuring the polarization of one photon of the early pair, and that it proves that quantum non-locality applies not only to space but also to time.

In three independent experiments in 2013 it was shown that classically-communicated separable quantum states can be used to carry entangled states. The first loophole-free Bell test was held in TU Delft in 2015 confirming the violation of Bell inequality.

In August 2014, Brazilian researcher Gabriela Barreto Lemos and team were able to "take pictures" of objects using photons that had not interacted with the subjects, but were entangled with photons that did interact with such objects. Lemos, from the University of Vienna, is confident that this new quantum imaging technique could find application where low light imaging is imperative, in fields like biological or medical imaging.

Mystery of Time

There have been suggestions to look at the concept of time as an emergent phenomenon that is a side effect of quantum entanglement. In other words, time is an entanglement phenomenon, which places all equal clock readings (of correctly prepared clocks, or of any objects usable as clocks) into the same history. This was first fully theorized by Don Page and William Wootters in 1983. The Wheeler–DeWitt equation that combines general relativity and quantum mechanics – by leaving out time altogether – was introduced in the 1960s and it was taken up again in 1983, when the theorists Don Page and William Wootters made a solution based on the quantum phenomenon of entanglement. Page and Wootters argued that entanglement can be used to measure time.

In 2013, at the Istituto Nazionale di Ricerca Metrologica (INRIM) in Turin, Italy, researchers performed the first experimental test of Page and Wootters' ideas. Their result has been interpreted to confirm that time is an emergent phenomenon for internal observers but absent for external observers of the universe just as the Wheeler-DeWitt equation predicts.

Source for the Arrow of Time

Physicist Seth Lloyd says that quantum uncertainty gives rise to *entanglement*, the putative source of the arrow of time. According to Lloyd; "The arrow of time is an arrow of increasing correlations." The approach to entanglement would be from the perspective of the causal arrow of time, with the assumption that the cause of the measurement of one particle determines the effect of the result of the other particle's measurement.

Quantum Mechanical Framework

The following subsections are for those with a good working knowledge of the formal, mathematical description of quantum mechanics, including familiarity with the formalism and theoretical framework developed in the articles: bra–ket notation and mathematical formulation of quantum mechanics.

Pure States

Consider two noninteracting systems A and B, with respective Hilbert spaces H_A and H_B. The Hilbert space of the composite system is the tensor product,

$$H_A \otimes H_B.$$

If the first system is in state $|\psi\rangle_A$ and the second in state $|\phi\rangle_B$, the state of the composite system is,

$$|\psi\rangle_A \otimes |\phi\rangle_B.$$

States of the composite system that can be represented in this form are called *separable states*, or *product states*.

Not all states are separable states (and thus product states). Fix a basis $\{|i\rangle_A\}$ for H_A and a basis $\{|j\rangle_B\}$ for H_B. The most general state in $H_A \otimes H_B$ is of the form,

$$|\psi\rangle_{AB} = \sum_{i,j} c_{ij} |i\rangle_A \otimes |j\rangle_B.$$

This state is separable if there exist vectors $[c_i^A],[c_j^B]$ so that $c_{ij} = c_i^A c_j^B$, yielding $|\psi\rangle_A = \sum_i c_i^A |i\rangle_A$ and $|\phi\rangle_B = \sum_j c_j^B |j\rangle_B$. It is inseparable if for any vectors $[c_i^A],[c_j^B]$ at least for one pair of coordinates c_i^A, c_j^B we have $c_{ij} \neq c_i^A c_j^B$. If a state is inseparable, it is called an *entangled state*.

For example, given two basis vectors $\{|0\rangle_A, |1\rangle_A\}$ of H_A and two basis vectors $\{|0\rangle_B, |1\rangle_B\}$ of H_B, the following is an entangled state:

$$\tfrac{1}{\sqrt{2}}\left(|0\rangle_A \otimes |1\rangle_B - |1\rangle_A \otimes |0\rangle_B\right).$$

If the composite system is in this state, it is impossible to attribute to either system A or system B a definite pure state. Another way to say this is that while the von Neumann entropyof the whole state is zero (as it is for any pure state), the entropy of the subsystems is greater than zero. In this sense, the systems are "entangled". This has specific empirical ramifications for interferometry. It is worthwhile to note that the above example is one of four Bell states, which are (maximally) entangled pure states (pure states of the $H_A \otimes H_B$space, but which cannot be separated into pure states of each H_A and H_B).

Now suppose Alice is an observer for system A, and Bob is an observer for system B. If in the entangled state given above Alice makes a measurement in the $\{|0\rangle, |1\rangle\}$ eigenbasis of A, there are two possible outcomes, occurring with equal probability:

1. Alice measures 0, and the state of the system collapses to $|0\rangle_A |1\rangle_B$.
2. Alice measures 1, and the state of the system collapses to $|1\rangle_A |0\rangle_B$.

If the former occurs, then any subsequent measurement performed by Bob, in the same basis, will always return 1. If the latter occurs, (Alice measures 1) then Bob's measurement will return 0 with certainty. Thus, system B has been altered by Alice performing a local measurement on system A. This remains true even if the systems A and B are spatially separated. This is the foundation of the EPR paradox.

The outcome of Alice's measurement is random. Alice cannot decide which state to collapse the composite system into, and therefore cannot transmit information to Bob by acting on her system. Causality is thus preserved, in this particular scheme.

Ensembles

A state of a quantum system is given by a unit vector in a Hilbert space. More generally, if one has less information about the system, then one calls it an *ensemble* and describes it by a density matrix, which is a positive-semidefinite matrix, or a trace class when the state space is infinite-dimensional, and has trace 1. Again, by the spectral theorem, such a matrix takes the general form:

$$\rho = \sum_i w_i |\alpha_i\rangle\langle\alpha_i|,$$

where the w_i are positive-valued probabilities (they sum up to 1), the vectors α_i are unit vectors, and in the infinite-dimensional case, we would take the closure of such states in the trace norm. We can interpret ρ as representing an ensemble where w_i is the proportion of the ensemble whose states are $|\alpha_i\rangle$. When a mixed state has rank 1, it therefore describes a *pure ensemble*. When there is less than total information about the state of a quantum system we need density matrices to represent the state.

Experimentally, a mixed ensemble might be realized as follows. Consider a "black box" apparatus that spits electrons towards an observer. The electrons' Hilbert spaces are identical. The apparatus might produce electrons that are all in the same state; in this case, the electrons received by the observer are then a pure ensemble. However, the apparatus could produce electrons in different states. For example, it could produce two populations of electrons: one with state $|\mathbf{z}+\rangle$ with spins aligned in the positive z direction, and the other with state $|\mathbf{y}-\rangle$ with spins aligned in the negative y direction. Generally, this is a mixed ensemble, as there can be any number of populations, each corresponding to a different state.

Following the definition above, for a bipartite composite system, mixed states are just density matrices on $H_A \otimes H_B$. That is, it has the general form,

$$\rho = \sum_i w_i \left[\sum_j \bar{c}_{ij} (|\alpha_{ij}\rangle \otimes |\beta_{ij}\rangle) \right] \otimes \left[\sum_k c_{ik} (\langle\alpha_{ik}| \otimes \langle\beta_{ik}|) \right]$$

where the w_i are positively valued probabilities, $\sum_j |c_{ij}|^2 = 1$, and the vectors are unit vectors. This is self-adjoint and positive and has trace 1.

Extending the definition of separability from the pure case, we say that a mixed state is separable if it can be written as,

$$\rho = \sum_i w_i \rho_i^A \otimes \rho_i^B,$$

where the w_i are positively valued probabilities and the ρ_i^A's and ρ_i^B's are themselves mixed states (density operators) on the subsystems A and B respectively. In other words, a state is separable if it is a probability distribution over uncorrelated states, or product states. By writing the density matrices as sums of pure ensembles and expanding, we may assume without loss of generality that ρ_i^A and ρ_i^B are themselves pure ensembles. A state is then said to be *entangled* if it is not separable.

In general, finding out whether or not a mixed state is entangled is considered difficult. The general bipartite case has been shown to be NP-hard. For the 2×2 and 2×3 cases, a necessary and sufficient criterion for separability is given by the famous Positive Partial Transpose (PPT) condition.

Reduced Density Matrices

The idea of a reduced density matrix was introduced by Paul Dirac in 1930. Consider as above systems A and B each with a Hilbert space H_A, H_B. Let the state of the composite system be,

$$|\Psi\rangle \in H_A \otimes H_B.$$

In general there is no way to associate a pure state to the component system A. However, it still is possible to associate a density matrix. Let,

$$\rho_T = |\Psi\rangle\langle\Psi|.$$

which is the projection operator onto this state. The state of A is the partial trace of ρ_T over the basis of system B:

$$\rho_A \stackrel{\text{def}}{=} \sum_j \langle j|_B \left(|\Psi\rangle\langle\Psi|\right)|j\rangle_B = \mathrm{Tr}_B\ \rho_T.$$

ρ_A is sometimes called the reduced density matrix of ρ on subsystem A. Colloquially, we "trace out" system B to obtain the reduced density matrix on A.

For example, the reduced density matrix of A for the entangled state

$$\tfrac{1}{\sqrt{2}}\left(|0\rangle_A \otimes |1\rangle_B - |1\rangle_A \otimes |0\rangle_B\right),$$

discussed above is,

$$\rho_A = \tfrac{1}{2}\left(|0\rangle_A\langle 0|_A + |1\rangle_A\langle 1|_A\right)$$

This demonstrates that, as expected, the reduced density matrix for an entangled pure ensemble is a mixed ensemble. Also not surprisingly, the density matrix of A for the pure product state $|\psi\rangle_A \otimes |\phi\rangle_B$ discussed above is,

$$\rho_A = |\psi\rangle_A\langle\psi|_A.$$

In general, a bipartite pure state ρ is entangled if and only if its reduced states are mixed rather than pure.

Two Applications that use them

Reduced density matrices were explicitly calculated in different spin chains with unique ground state. An example is the one-dimensional AKLT spin chain: the ground state can be divided into a block and an environment. The reduced density matrix of the block is proportional to a projector to a degenerate ground state of another Hamiltonian.

The reduced density matrix also was evaluated for XY spin chains, where it has full rank. It was proved that in the thermodynamic limit, the spectrum of the reduced density matrix of a large block of spins is an exact geometric sequence in this case.

Entanglement as a Resource

In quantum information theory, entangled states is considered a *resource*, i.e., something costly to produce and that allows to implement valuable transformations. The setting in which this perspective is most evident is that of "distant labs", i.e., two quantum systems labeled "A" and "B" on each of which arbitrary quantum operations can be performed, but which do not interact with each other quantum mechanically. The only interaction allowed is the exchange of classical information, which combined with the most general *local*quantum operations gives rise to the class of operations called *LOCC* (local operations and classical communication). These operations do not allow to produce entangled states between the systems A and B. But if A and B are provided with a supply of entangled states, then these, together with LOCC operations can enable a larger class of transformations. For example, an interaction between a qubit of A and a qubit of B can be realized by first teleporting A's qubit to B, then letting it interact with B's qubit (which is now a LOCC operation, since both qubits are in B's lab) and then teleporting the qubit back to A. Note that two maximally entangled states of two qubits are used up in this process. Thus entangled states are a resource that enables the realization of quantum interactions (or of quantum channels) in a setting where only LOCC are available, but they is consumed in the process. There are other applications where entanglement can be seen as a resource, e.g., private communication or distinguishing quantum states.

Classification of Entanglement

Not all quantum states are equally valuable as a resource. To quantify this value, different entanglement measures can be used, that assign a numerical value to each quantum state. However, it is often interesting to settle for a coarser way to compare quantum states. This gives rise to different classification schemes. Most entanglement classes are defined based on whether states can be converted to other states using LOCC or a subclass of these operations. The smaller the set of allowed operations, the finer the classification. Important examples are:

- If two states can be transformed into each other by a local unitary operation, they are said to be in the same *LU class*. This is the finest of the usually considered classes. Two states in the same LU class have the same value for entanglement measures and the same value as a resource in the distant-labs setting. There is an infinite number of different LU classes (even in the simplest case of two qubits in a pure state).
- If two states can be transformed into each other by local operations including measurements with probability larger than 0, they are said to be in the same *SLOCC class*("stochastic LOCC"). Qualitatively, two states ρ_1 and ρ_2 in the same SLOCC class are equally powerful (since I can transform one into the other and then do whatever it allows me to do), but since the transformations $\rho_1 \to \rho_2$ and $\rho_2 \to \rho_1$ may succeed with different probability, they are no longer equally valuable. E.g., for two pure qubits there are only two SLOCC classes: the entangled states (which contains both the (maximally entangled) Bell states and weakly entangled states like $|00\rangle + 0.01|11\rangle$) and the separable ones (i.e., product states like $|00\rangle$).

- Instead of considering transformations of single copies of a state (like $\rho_1 \to \rho_2$) one can define classes based on the possibility of multi-copy transformations. E.g., there are examples when $\rho_1 \to \rho_2$ is impossible by LOCC, but $\rho_1 \otimes \rho_1 \to \rho_2$ *is* possible. A very important (and very coarse) classification is based on the property whether it is possible to transform an *arbitrarily large* number of copies of a state ρ into a at least on pure entangled state. States that have this property are called *distillable*. These states are the most useful quantum states since, given enough of them, they can be transformed (with local operations) into *any* entangled state and hence allow for all possible uses. It came initially as a surprise that not all entangled states are distillable, those that are not are called *bound entangled*.

A different entanglement classification is based on what the quantum correlations present in a state allow A and B to do: one distinguishes three subsets of entangled states: (1) the *non-local states*, which produce correlations that cannot be explained by a local hidden variable model and thus violate a Bell inequality, (2) the *steerable states* that contain sufficient correlations for A to modify ("steer") by local measurements the conditional reduced state of B in such a way, that A can prove to B that the state they possess is indeed entangled, and finally (3) those entangled state that are neither non-local nor steerable. All three sets are non-empty.

Entropy

In this, the entropy of a mixed state is discussed as well as how it can be viewed as a measure of quantum entanglement.

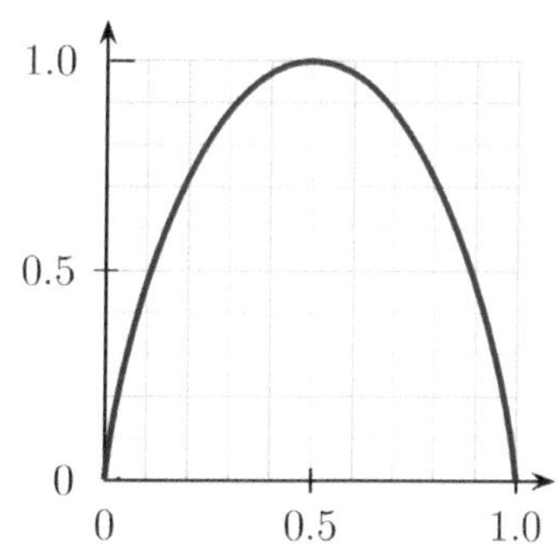

The plot of von Neumann entropy vs Eigenvalue for a bipartite 2-level pure state. When the eigenvalue has value .5, von Neumann entropy is at a maximum, corresponding to maximum entanglement

In classical information theory *H*, the Shannon entropy, is associated to a probability distribution, $p_1, \cdots, p_n$, in the following way:

$$H(p_1, \cdots, p_n) = -\sum_i p_i \log_2 p_i.$$

Since a mixed state ρ is a probability distribution over an ensemble, this leads naturally to the definition of the von Neumann entropy:

$$S(\rho) = -\mathrm{Tr}(\rho \log_2 \rho).$$

In general, one uses the Borel functional calculus to calculate a non-polynomial function such as $\log_2(\rho)$. If the nonnegative operator ρ acts on a finite-dimensional Hilbert space and has eigenvalues

$\lambda_1,\cdots,\lambda_n$, $\log_2(\rho)$ turns out to be nothing more than the operator with the same eigenvectors, but the eigenvalues $log_2(\lambda_1),\cdots,log_2(\lambda_n)$. The Shannon entropy is then:

$$S(\rho) = -\mathrm{Tr}(\rho\, log_2\, \rho) = -\sum_i \lambda_i\, log_2\, \lambda_i.$$

Since an event of probability 0 should not contribute to the entropy, and given that,

$$\lim_{p\to 0} p \log p = 0,$$

the convention 0 log(0) = 0 is adopted. This extends to the infinite-dimensional case as well: if ρ has spectral resolution,

$$\rho = \int \lambda dP_\lambda$$

assume the same convention when calculating,

$$\rho\, log_2\, \rho = \int \lambda\, log_2\, \lambda dP_\lambda.$$

As in statistical mechanics, the more uncertainty (number of microstates) the system should possess, the larger the entropy. For example, the entropy of any pure state is zero, which is unsurprising since there is no uncertainty about a system in a pure state. The entropy of any of the two subsystems of the entangled state discussed above is second log (which can be shown to be the maximum entropy for 2 × 2 mixed states).

As a Measure of Entanglement

Entropy provides one tool that can be used to quantify entanglement, although other entanglement measures exist. If the overall system is pure, the entropy of one subsystem can be used to measure its degree of entanglement with the other subsystems.

For bipartite pure states, the von Neumann entropy of reduced states is the unique measure of entanglement in the sense that it is the only function on the family of states that satisfies certain axioms required of an entanglement measure.

It is a classical result that the Shannon entropy achieves its maximum at, and only at, the uniform probability distribution $\{1/n, ..., 1/n\}$. Therefore, a bipartite pure state $\rho \in H_A \otimes H_B$ is said to be a maximally entangled state if the reduced state of ρ is the diagonal matrix,

$$\begin{bmatrix} \frac{1}{n} & & \\ & \ddots & \\ & & \frac{1}{n} \end{bmatrix}.$$

For mixed states, the reduced von Neumann entropy is not the only reasonable entanglement measure.

As an aside, the information-theoretic definition is closely related to entropy in the sense of statistical mechanics (comparing the two definitions, we note that, in the present context, it is customary to set the Boltzmann constant $k = 1$). For example, by properties of the Borel functional calculus, we see that for any unitary operator U,

$$S(\rho) = S\left(U\rho U^{*}\right).$$

Indeed, without this property, the von Neumann entropy would not be well-defined.

In particular, U could be the time evolution operator of the system, i.e.,

$$U(t) = exp\left(\frac{-iHt}{\hbar}\right),$$

where H is the Hamiltonian of the system. Here the entropy is unchanged.

The reversibility of a process is associated with the resulting entropy change, i.e., a process is reversible if, and only if, it leaves the entropy of the system invariant. Therefore, the march of the arrow of time towards thermodynamic equilibrium is simply the growing spread of quantum entanglement. This provides a connection between quantum information theory and thermodynamics.

Rényi entropy also can be used as a measure of entanglement.

Entanglement Measures

Entanglement measures quantify the amount of entanglement in a (often viewed as a bipartite) quantum state. As aforementioned, entanglement entropy is the standard measure of entanglement for pure states (but no longer a measure of entanglement for mixed states). For mixed states, there are some entanglement measures in the literature and no single one is standard.

- Entanglement cost
- Distillable entanglement
- Entanglement of formation
- Relative entropy of entanglement
- Squashed entanglement
- Logarithmic negativity.

Most (but not all) of these entanglement measures reduce for pure states to entanglement entropy, and are difficult (NP-hard) to compute.

Quantum Field Theory

The Reeh-Schlieder theorem of quantum field theory is sometimes seen as an analogue of quantum entanglement.

Wave Function

The wave function is the most fundamental concept of quantum mechanics. It was first introduced into the theory by analogy (Schrödinger 1926); the behavior of microscopic particles likes wave, and thus a wave function is used to describe them. Schrödinger originally regarded the wave function as a description of real physical wave. But this view met serious objections and was soon replaced by Born's probability interpretation (Born 1926), which becomes the standard interpretation of the wave function today. According to this interpretation, the wave function is a probability amplitude, and the square of its absolute value represents the probability density for a particle to be measured in certain locations. However, the standard interpretation is still unsatisfying when applying to a fundamental theory because of resorting to measurement. In view of this problem, some alternative realistic interpretations of the wave function have been proposed and widely studied (Bohm 1952; Everett 1957; Nelson 1966; Ghirardi, Grassi and Benatti 1995).

There are in general two possible ways to interpret the wave function of a single quantum system in a realistic interpretation. One view is to take the wave function as a physical entity simultaneously distributing in space such as a field, and it is assumed by de Broglie-Bohm theory, many-worlds interpretation and dynamical collapse theories etc (de Broglie 1928; Bohm 1952; Everett 1957; Ghirardi, Grassi and Benatti 1995) . For example, in de Broglie-Bohm theory the wave function is generally taken as an objective physical field, called Ψ-field, though there are various views on exactly what field the wave function is. The other view is to take the wave function as a description of some kind of ergodic motion of a particle (or corpuscle), and it is assumed by stochastic interpretation etc. The essential difference between a field and the ergodic motion of a particle lies in the property of simultaneity. The field exists throughout space simultaneously, whereas the ergodic motion of a particle exists throughout space in an essentially local way; the particle is still in one position at each instant, and it is only during a time interval that the ergodic motion of the particle spreads throughout space.

Wave Function: Physical Field or Not

If the wave function is a physical field, then its mass and charge density will simultaneously distribute in space. This has two disaster consequences at least. One is that charge will not be quantized; the total charge inside a very small region can be much smaller than an elementary charge for a single quantum system. This obviously contradicts the common expectation that charge should be quantized. But maybe our expectation needs to be revised. So this consequence is not fatal for the field explanation of the wave function. The other is that the wave function will not satisfy the superposition principle. For example, for the wave function of a single electron, different spatial parts of the wave function will have gravitational and electrostatic interactions, as these parts have mass and charge simultaneously.

Interestingly, the so-called Schrödinger-Newton equation, which was proposed for other purposes (Diosi 1984; Penrose 1998), just describes the gravitational self-interaction of the wave function. The equation for a single quantum system can be written as,

$$i\hbar\frac{\partial\psi(x,t)}{\partial t}=-\frac{\hbar^2}{2m}\nabla^2\psi(x,t)-Gm^2\int\frac{|\psi(x',t)|^2}{|x-x'|}d^3x'\psi(x,t)+V\psi(x,t)$$

where m is the mass of the quantum system, V is an external potential, and G is Newton's gravitational constant. Much work has been done to study the mathematical properties of this interesting equation. Some experimental schemes have been also proposed to test its physical validity (Salzman and Carlip 2006). As we will see, although such gravitational self-interactions cannot yet be excluded by experiments, the existence of electrostatic self-interaction already contradicts experimental observations.

If there is also an electrostatic self-interaction, then the equation for a free quantum system with mass m and charge Q will be,

$$i\hbar\frac{\partial\psi(x,t)}{\partial t}=-\frac{\hbar^2}{2m}\nabla^2\psi(x,t)+\left(kQ^2-Gm^2\right)\int\frac{\left|\psi(x',t)\right|^2}{\left|x-x'\right|}d^3x'\psi(x,t)$$

where k is the Coulomb constant. Note that the gravitational self-interaction is an attractive force, while the electrostatic self-interaction is a repulsive force. It has been shown that the measure of the potential strength of a gravitational self-interaction is $\varepsilon^2=\left(\frac{4Gm^2}{\hbar c}\right)^2$ for a free particle with mass m (Salzman 2005). This quantity represents the strength of the influence of self-interaction on the normal evolution of the wave function; when 1 2 ε ≈ the influence will be significant. Similarly, for a free charged particle with charge Q, the measure of the potential strength of the electrostatic self-interaction is $\varepsilon^2=\left(\frac{4kQ^2}{\hbar c}\right)^2$. As a typical example, for a free electron with charge e, the potential strength of the electrostatic self-interaction will be $\varepsilon^2=\left(\frac{4kQ^2}{\hbar c}\right)^2\approx 1\times 10^{-3}$. This indicates that the electrostatic self-interaction will have significant influence on the evolution of the wave function of a free electron. If such an interaction indeed exists, it should have been detected by precise experiments on charged microscopic particles. As another example, consider the electron in the hydrogen atom. Since the potential of its electrostatic self-interaction is of the same order as the Coulomb potential produced by the nucleus, the energy levels of hydrogen atoms will be significantly different from those predicted by quantum mechanics and confirmed by experimental observations. Therefore, the electrostatic self-interaction cannot exist for the wave function of a charged quantum system. Since the field explanation of the wave function entails the existence of such electrostatic self-interactions, it cannot be right, i.e. the wave function cannot be a description of a physical field.

One may object to the above argument with the example of classical electromagnetic field. Electromagnetic field is a field, but it has no self-interaction. Thus a field does not require the existence of self-interaction. However, this is a common misunderstanding. The crux of the matter is that the non-existence of electromagnetic self-interaction results from the fact that electromagnetic field itself has no charge. If the electromagnetic field had charge, then there would also exist electromagnetic self-interaction due to the nature of field, namely the simultaneous existence of its properties in space. In fact, although electromagnetic field has no electromagnetic self-interaction, it does have gravitational self-interaction; the simultaneous existence of energy densities in different spatial locations for an electromagnetic field must generate a gravitational interaction, though the interaction is too weak to be detected by current technology.

One may further object that the superposition principle in quantum mechanics already prohibits the existence of the above self-interactions. But this is just the key point we use to argue against the field explanation of the wave function. Let's state the argument more explicitly. If the wave function of a charged quantum system is a physical field, then the different spatial parts of this field will have gravitational and electrostatic interactions. But the superposition principle in quantum mechanics, which has been verified within astonishing precision, does not permit the existence of the remarkable electrostatic self-interaction. Therefore, the field explanation of the wave function is already refuted by the superposition principle of quantum mechanics.

Classical Ergodic Models Fails

If the wave function is not a description of a physical field, then exactly what does the wave function describe? This naturally leads us to the second view that takes the wave function as a description of some sort of ergodic motion of particles. On this view, the effective mass and charge density are formed by time average of the motion of a charged particle, and they distribute in different locations at different moments. At every instant, there is only a localized particle with mass and charge. Thus there will not exist any self-interaction for the wave function, and this view can be consistent with quantum mechanics and experimental observations. In fact, if the mass and charge density does not exist in different regions simultaneously as the field explanation holds, they can only be formed by the ergodic motion of a particle. As a result, the wave function must be a description of some sort of ergodic motion of particles.

There are indeed some realistic interpretations of quantum mechanics that attempt to explain the wave function in terms of the ergodic motion of particles. A well-known example is the stochastic interpretation of quantum mechanics (e.g. Nelson 1966). Nelson (1966) derived the Schrödinger equation from Newtonian mechanics via the hypothesis that every particle of mass m is subject to a Brownian motion with diffusion coefficient $\hbar/2m$ and no friction. In more technical terms, the quantum mechanical process is claimed to be equivalent to a classical Markovian diffusion process. On this interpretation, particles have continuous trajectories but no velocities, and the wave function is a statistical average description of their ergodic motion.

However, it has been pointed out that the classical stochastic interpretations are inconsistent with quantum mechanics (Glabert, Hänggi and Talkner 1979; Wallstrom 1994). Glabert, Hänggi and Talkner (1979) argued that the Schrödinger equation is not equivalent to a Markovian process, and the various correlation functions used in quantum mechanics do not have the properties of the correlations of a classical stochastic process. Wallstrom (1994) further showed that one must add by hand a quantization condition, as in the old quantum theory, in order to recover the Schrödinger equation, and thus the Schrödinger equation and the Madelung hydrodynamic equations are not equivalent. In fact, Nelson (2005) also showed that there is an empirical difference between the predictions of quantum mechanics and his stochastic mechanics when considering quantum entanglement and nonlocality. For example, for two widely-separated but entangled harmonic oscillators, the two theories predict totally different statistics; stochastic mechanics predicts that measurements of the position of the first one at time T (oscillation period) and the position of the second one at time 0 do not interfere with each other, while quantum mechanics predicts that there exists a strong correlation between them.

In addition, it can be generally argued that the classical ergodic models that assume continuous motion of particles cannot be consistent with quantum mechanics. In order to see this let's examine whether the continuous motion of particles can generate the charge density $Q|\psi(x,t)|^2$ at the level of time average. Consider an electron in a one-dimensional box in an energy eigenstate such as the first excited state. Its wave function has a node at the center of the box, where its charge density is zero. The electron performs a very fast motion in the box. At a particular time the charge density is either zero (if the electron is not there) or singular (if the electron is inside the infinitesimally small region including the space point in question). But during a time interval, the motion of the electron will generate a charge density cloud with an effective charge density. The question is whether this density can assume the same form as $e|\psi(x)|^2$. Since the effective charge density is proportional to the amount of time the electron spends in a given position, the electron must be in the left half of the box half of the time and in the right half of the box half of the time. But it can spend no time at the center of the box where the effective charge density is zero; in other words, it must move at infinite velocity at the center. Certainly, the appearance of infinite velocity or velocity faster than light may be not a fatal problem, as our discussion is entirely in the context of non-relativistic quantum mechanics and especially the infinite potential assumed in the example is also an ideal situation. However, it is hard to understand why the electron speeds up at the node and where the infinite energy required for the acceleration comes from. Moreover, the sudden acceleration of the electron near the node will result in large radiation, which is inconsistent with both the predictions of quantum mechanics and experimental observations. Maybe one can also assume in an ad hoc way that the accelerating electron does not radiate here in order to make the model be consistent with quantum mechanics and experimental observations.

Let's further consider a superposition of two energy eigenstate respectively limited in two widely-separated boxes. In this example, even if one assumes that the electron can move with an infinite velocity (e.g. at the nodes), it cannot move from one box to another due to the limitation of box walls. Therefore, any sort of continuous motion cannot generate the charge density $e|\psi(x)|^2$ at the level of time average. One may still object that this is merely an artificial result of the idealization of infinite potential. But even in this ideal example, the model should also be able to generate the charge density by means of some sort of ergodic motion of the electron. In fact, there is a very similar situation in double-slit experiment. The wave function of a single electron passes through two channels that are well separated in space. The wave function disappears outside the channels for all practical purposes, and the electron can only move inside the channels (otherwise the electron will be detected outside the channels, which contradicts experimental observations). Again, a classical ergodic model cannot explain this experiment.

As we think, there is a general objection to all classical ergodic models. Any classical ergodic model will inevitably introduce a finite ergodic time parameter, which is needed to generate the effective mass and charge density, because it must take a finite time for the particle to continuously move throughout all regions where the wave function is not zero. However, it can be argued that no finite time scale is permitted to exist for the ergodic motion. First of all, the existence of a finite time scale, denoted by T_c, is inconsistent with the standard quantum theory, as there is no such a time constant in the theory. Next, if there exists a time scale T_c, then whenthe measuring time T of protective measurement is shorter than T_c (i.e. $T < T_c$), the measurement result will be not the expectation value of a variable such as charge density, as no whole time average can be obtained.

This also contradicts the prediction of protective measurement. As an extreme example, consider a spatial superposition state $\psi_L + \psi_R$, where ψ_L and ψ_R are Gaussian wave packets and their centers are well separated in space. When $T < T_c$, the particle has no enough time to move throughout the whole regions including both L and R. Then the result of a protective position measurement will be not the expectation value of $\psi_L + \psi_R$, but the eigenvalue corresponding to ψ_L or ψ_R. Moreover, the results distribution is also different from that predicted by protective measurement. When $T < T_c$, the distribution of position measurement results will concentrate near L and R, while according to protective measurement, the distribution should concentrate near the midpoint between L and R. In fact, for protective measurement, during any period of time the pointer of the measuring device always shifts by an amount proportional to the expectation value of the measured variable, rather than to one of its eigenvalues. Thus the expectation value can be associated with any short period of time. Certainly, pointer shifts during short time intervals are practically unobservable since they are much smaller than the uncertainty, and only when the total shift accumulated during the whole period of measurement is much larger than the width of the initial distribution it becomes observable.

Therefore, we conclude that the continuous ergodic motion of particles cannot generate the effective mass and charge density measurable by protective measurement, and the classical ergodic models cannot be consistent with quantum mechanics. As a result, the wave function cannot be a description of the continuous ergodic motion of particles.

The Wave Function as a Description of Quantum Motion of Particles

The failure of classical ergodic models does not exclude all possible ergodic motion of particles. In this, we will argue that another different kind of motion – random discontinuous motion can naturally generate the effective mass and charge density measurable by protective measurement, and what the wave function describes is probably such quantum motion of particles, which is essentially discontinuous and random.

If the motion of a particle is not continuous but discontinuous and random, and the probability density of the particle being in certain positions is proportional to the square of the absolute value of its wave function at every instant, then the particle can readily move throughout all possible regions where the wave function is nonzero during an arbitrarily short time interval near a given instant. This will solve the problems plagued by the classical ergodic models. The discontinuous ergodic motion requires no existence of a finite time scale. A particle undergoing discontinuous motion can also move from one region to another spatially separated region, no matter whether there is an infinite potential wall between them. Besides, discontinuous motion can also solve the problems of infinite velocity and accelerating radiation. The reason is that no classical velocity and acceleration can be defined for discontinuous motion, and energy and momentum will require new definition and new understanding as in quantum mechanics. Thus it seems that the discontinuous ergodic motion of particles can in principle generate the effective mass and charge density measurable by protective measurement, and thus the wave function is probably a description of such random discontinuous motion of particles.

In some sense, the above interpretation of the wave function seems to be an inevitable consequence of protective measurement. According to protective measurement, a charged quantum system has effective mass and charge density distributing in space, proportional to the square of the absolute value of its wave function. There are two possible ways to explain the existence of the mass and charge density; one is to take the wave function as a description of some kind of physical field, and the mass and charge density of this field exists throughout space simultaneously, the other is to take the wave function as a description of some sort of ergodic motion of particles, and the effective mass and charge density formed by such motion exists throughout space in an essentially local way. The first view has been rejected because it entails the existence of a remarkable electrostatic self-interaction that contradicts experimental observations. Thus the wave function can only be a description of some sort of ergodic motion of particles. Since the classical ergodic models have also been excluded, the ergodic motion of particles cannot be continuous and must be essentially discontinuous. Besides, when considering the randomness of the results of conventional quantum measurement, such motion must be also random. Therefore, what the wave function describes can only be random discontinuous motion of particles.

In fact, by assuming the wave function is a (complete) description for the motion of particles, we can reach the random discontinuous motion in a more direct way, independent of the analysis of protective measurement. If the wave function $\psi(x,t)$ is a description of the state of motion for a single particle, then the quantity $|\psi(x,t)|^2 dx$ not only gives the probability of the particle being found in an infinitesimal space interval dx near position x at instant t (as in standard quantum mechanics), but also gives the objective probability of the particle being there. This accords with the well-accepted assumption that the probability distribution of the measurement outcomes of a property is the same as the actual distribution of the property in the measured state. Then at instant *t* the particle may appear in any location where the probability density $|\psi(x,t)|^2$ is nonzero, and during an infinitesimal time interval near instant t, the particle will move throughout the whole space where the wave function $\psi(x,t)$ spreads, though it is still in one position at each instant. Moreover, the density distribution of its positions is equal to the probability density $|\psi(x,t)|^2$. Obviously, this kind of motion is essentially random and discontinuous.

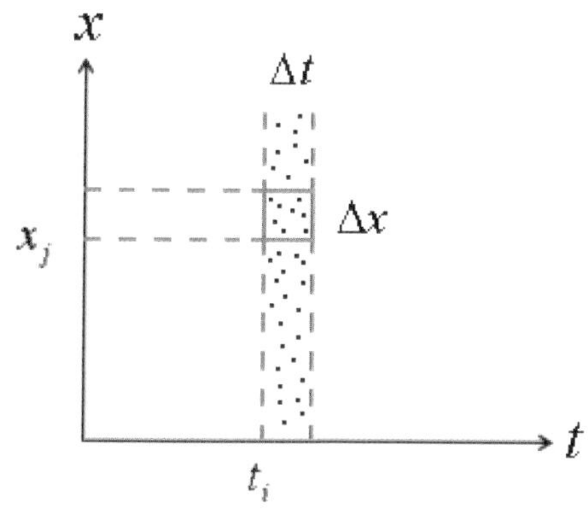

The description of RDM of a single particle

The strict mathematical description of random discontinuous motion (RDM henceforth) can be obtained by using the measure theory. Consider the motion state of a single particle in finite intervals Δt and Δx near a space-time point (t_i, x_j) as shown in figure above. For RDM, the position of the particle forms a random discontinuous point set in the whole space for the time interval Δt near the instant t_i. Accordingly, there is a local discontinuous point set in the space interval Δx near the position x_j. The local discontinuous point set represents the motion state of the particle in the finite intervals Δt and Δx near the space-time point (t_i, x_j). We study its projection in the t-axis, namely the corresponding dense instant set in the time interval Δt . Let W be the discontinuous trajectory or world-set of the particle and Q be the square region $\left[x_j, x_j+\Delta x\right]\times\left[t_i, t_i+\Delta t\right]$. The dense instant set can be denoted by $\pi_t(W\cap Q)\subset\Re$, where π_t is the projection on the t-axis. According to the measure theory, we can define the Lebesgue measure:

$$M_{\text{Äx,Ät}}\left(x_j, t_i\right)=\int_{\pi_t(W\cap Q)\subset\Re} dt$$

Since the sum of the measures of all such dense instant sets in the time interval Δt is equal to the length of the continuous time interval Δt, we have:

$$\sum_j M_{\text{Äx,Ät}}\left(x_j, t_i\right)=\text{Ät}$$

Then we can define the measure density:

$$\rho(x,t)=\lim_{\substack{\text{Äx}\to 0\\ \text{Ät}\to 0}} M_{\text{Äx,Ät}}\left(x_j, t_i\right)/(\text{Äx}\cdot\text{Ät})$$

The limit exists for a random discontinuous point set. This provides a strict mathematical description of the point distribution situation for the above local discontinuous point set. We call this measure density position measure density.

Since the local discontinuous point set represents the motion state of the particle, the position measure density ρ(x, t) will be a descriptive quantity for the RDM of a single particle. It represents the relative frequency of the particle appearing in an infinitesimal space interval dx near position x during an infinitesimal interval dt near instant t . From equation above we can see that ρ(x, t) satisfies the normalization relation, namely $\int_{-\infty}^{+\infty}\rho(x,t)dx=1$. Furthermore, we can define the position measure flux density j(x, t) through the relation j(x, t) = ρ(x, t)v(x, t), where v(x, t) is the velocity of the local discontinuous point set. It describes the change of the position measure density with time. Due to the conservation of measure, ρ(x, t) and j(x, t) satisfy the following equation:

$$\frac{\partial\rho(x,t)}{\partial t}+\frac{\partial j(x,t)}{\partial x}=0$$

exists a real number δ > 0 such that whenever a point z ∈ A has distance less than δ to x, the point f (z)∈R has distance less than ε to f (x) .

The position measure density ρ(x, t) and the position measure flux density j(x, t) provide a complete description for the RDM of a single particle.

It is very natural to extend the description of the motion of a single particle to the motion of many particles. For the RDM state of N particles, we can define a joint position measure density $\rho(x_1, x_2, \ldots x_N, t)$. This represents the relative probability of the situation in which particle 1 is in position x_1, particle 2 is in position x_2, ..., and particle N is in position x_N. In a similar way, we can define the joint position measure flux density $j(x_1, x_2, \ldots x_N, t)$. It satisfies the joint measure conservation equation:

$$\frac{\partial\rho(x_1,x_2,...x_N,t)}{\partial t}+\sum_{i=1}^{N}\frac{\partial j(x_1,x_2,...x_N,t)}{\partial x_i}=0$$

When these N particles are independent, the joint position measure density $\rho(x_1,x_2,...x_N,t)$ can be reduced to the direct product of the position measure density of each particle, namely $\rho(x_1,x_2,...x_N,t)=\prod_{i=1}^{N}\rho(x_i,t)$. It is worth noting that the joint position measure density $\rho(x_1,x_2,...x_N,t)$ and the joint position measure flux density $j(x_1,x_2,...x_N,t)$ are not defined in the three-dimensional real space, but defined in the 3N-dimensional configuration space. These two quantities can further constitute the wave function, and the many-body wave functions thus defined also live on the 3N-dimensional configuration space.

With respect to the RDM of a particle, the motion of the particle is completely discontinuous and random. The probability density for the particle to appear at position x at instant t is $\rho(x, t)$. There is no evolution law for the position state of a particle, and the trajectory function x(t) is random and discontinuous at every instant. However, the discontinuity of RDM is absorbed into the motion state of a particle, which is defined during an infinitesimal time interval, by the descriptive quantities of position measure density $\rho(x, t)$ and position measure flux density $j(x, t)$. Therefore, the evolution law for the motion state of a particle will contain no discontinuities and can be a continuous equation.

By assuming that the nonrelativistic evolution equation of RDM is the Schrödinger equation, the wave function $\psi(x,t)$ can be uniquely expressed by the position measure density $\rho(x, t)$ and the position measure flux density $j(x, t)$:

$$\psi(x,t)=\sqrt{\rho(x,t)e^{iS(x,t)/h}}$$

where $S(x,t)=m\int_{-\infty}^{x}\frac{j(x,t)}{\rho(x,t)}dx$. Since $\rho(x, t)$ and $j(x, t)$ provide a complete description of the RDM of a single particle, the wave function $\psi(x,t)$ also provides a complete description of the RDM of a single particle.

The new interpretation of the wave function in terms of RDM of particles can be taken as a natural realistic alternative to the orthodox view. According to the standard probability interpretation of the wave function, the square of the absolute value of a N-particle wave function, which can be denoted by $\left|\ (x_1,x_2,...x_N,t)\right| dx_1,dx_2,...dx_N$, gives the probability of particle 1 being found in the infinitesimal interval dx_1 near position x_1 and particle 2 being found in the infinitesimal interval 2 dx near position x_2, ..., and particle N being found in the infinitesimal interval dx_N near position x_N. By contrast, according to the new interpretation, the square of the absolute value of the wave function not only gives the probability of a particle being found in certain locations, but also gives the objective probability of the particle being there. For example, $\left|\psi(x_1,x_2,...x_N,t)\right|^2 dx_1,dx_2,...dx_N$ also represents the objective probability of particle 1 being in the infinitesimal interval dx_1 near position x_1 and particle 2 being in the infinitesimal interval dx_2 near position x_2, ..., and particle N being in the infinitesimal interval dx_N near position x_N. Certainly, the transition process from "being" to "being found", which is closely related to the notorious quantum measurement problem, also needs to be explained.

Wave Particle Duality

Wave-particle duality is the concept that all matter exhibits the properties of both waves and particles. Central to quantum mechanics, the idea that light and matter can be treated as both wave-like and particle-like properties has been verified experimentally for subatomic particles, light, and even atoms and molecules. For example, matter can interfere with itself and also diffract, showing wave-like properties. And yet the photoelectric effect is clear evidence that light consists of discrete particle-like packets of energy, called photons. Seemingly contradictory, wave-particle duality implies that matter and light are both particles *and* waves. In reality, they are both and neither. The concepts of waves and particles, while familiar, are simply insufficient to accurately describe the true nature of matter.

Einstein had shown that the momentum of a photon is,

$$p = \frac{h}{\lambda}$$

This can be easily shown as follows. Assuming $E = h\nu$ for a photon and $\lambda\nu = c$ for an electromagnetic wave, we obtain,

$$E = \frac{hc}{\lambda}$$

Now we use Einstein's relativity result $E = mc^2$ to find

$$\lambda = \frac{h}{mc}$$

which is equivalent to equation $p = \frac{h}{\lambda}$. Note that m refers to the relativistic mass, not the rest mass, since the rest mass of a photon is zero. Since light can behave both as a wave (it can be diffracted, and it has a wavelength), and as a particle (it contains packets of energy $h\nu$), de Broglie reasoned in 1924 that matter also can exhibit this *wave-particle duality*. He further reasoned that matter would obey the same equation $p = \frac{h}{\lambda}$ as light. In 1927, Davisson and Germer observed diffraction patterns by bombarding metals with electrons, confirming de Broglie's proposition.

de Broglie's equation offers a justification for Bohr's assumption. If we think of an electron as a wave, then for the electron orbit to be stable the wave must complete an integral number of wavelengths during its orbit. Otherwise, it would interfere destructively with itself. This condition may be written as,

$$2\pi r = n\lambda$$

If we use the de Broglie relation, this can be rewritten as,

$$mvr = n\hbar$$

which is identical to Bohr's equation.

Although de Broglie's equation justifies Bohr's quantization assumption, it also demonstrates a deficiency of Bohr's model. Heisenberg showed that the wave-particle duality leads to the famous uncertainty principle,

$$\Delta x \Delta p \approx h$$

One result of the uncertainty principle is that if the orbital radius of an electron in an atom r is known exactly, then the angular momentum must be completely unknown. The problem with Bohr's model is that it specifies r exactly and it also specifies that the orbital angular momentum must be an integral multiple of $\hbar$. Thus the stage was set for a new quantum theory which was consistent with the uncertainty principle.

References

- Friedlander, Gerhart; Kennedy, Joseph E.; Miller, Julian Malcolm (1964). Nuclear and Radiochemistry (2nd ed.). New York: John Wiley & Sons. pp. 225–7. ISBN 978-0-471-86255-0
- Eddi, A.; Fort, E.; Moisy, F.; Couder, Y. (16 June 2009). "Unpredictable Tunneling of a Classical Wave-Particle Association" (PDF). Physical Review Letters. 102 (24): 240401. Bibcode:2009PhRvL.102x0401E. doi:10.1103/PhysRevLett.102.240401. PMID 19658983. Retrieved 1 May 2016
- Quantum-mechanics-explanation-33816: livescience.com, Retrieved 11 April 2018
- Jump up^ Laloe, Franck (2001), "Do We Really Understand Quantum Mechanics", American Journal of Physics, 69: 655–701, arXiv:quant-ph/0209123, Bibcode:2001AmJPh..69..655L, doi:10.1119/1.1356698
- Matta, Cherif F. (2014). Quantum Biochemistry: Electronic Structure and Biological Activity. Weinheim: Wiley-VCH. ISBN 978-3-527-62922-0
- Quantum-entanglement: sciencedaily.com, Retrieved 31 March 2018
- Sinha Roy, Soudip (2017-12-25). Generalized Quantum Tunneling Effect and Ultimate Equations for Switching Time and Cell to Cell Power Dissipation Approximation in QCA Devices. doi:10.13140/rg.2.2.23039.71849
- Mandelbaum, Ryan F. (11 May 2018). "100,000 Video Game Players Helped Scientists Prove Einstein Wrong". Gizmodo. Retrieved 12 May 2018
- Quantum-entanglement, epr-paradox-and-stern-gerlach-experiments: brilliant.org, Retrieved 18 May 2018
- Jump up^ Sanctuary, B. C (2006). "Correlations in Entangled States". International Journal of Modern Physics B. 20 (11n13): 1496–1503. arXiv:quant-ph/0508238. doi:10.1142/S0217979206034078
- Majumdar, Rabi (2011). Quantum Mechanics: In Physics and Chemistry with Applications to Bioloty. Newi: PHI Learning. ISBN 9788120343047

Relativistic Mechanics

An important area of physics is relativistic mechanics, which studies mechanics that is compatible with general and special relativity. This chapter has been carefully written to provide an easy understanding of the important concepts of relativistic mechanics, special relativity and general relativity, and includes vital topics such as time dilation, Lorentz transformation, relativistic space-time, relativistic mass, momentum and energy, etc.

Relativistic mechanics, science concerned with the motion of bodies whose relative velocities approach the speed of light c, or whose kinetic energies are comparable with the product of their masses m and the square of the velocity of light, or mc^2. Such bodies are said to be relativistic, and when their motion is studied, it is necessary to take into account Einstein's special theory of relativity. As long as gravitational effects can be ignored, which is true so long as gravitational potential energydifferences are small compared with mc^2, the effects of Einstein's general theory of relativity may be safely ignored.

The bodies concerned may be sufficiently small that one may ignore their internal structure and size and regard them as point particles, in which case one speaks of relativistic point-particle mechanics; or one may need to take into account their internal structure, in which case one speaks of relativistic continuum mechanics. This part is concerned only with relativistic point-particle mechanics. It is also assumed that quantum mechanical effects are unimportant, otherwise relativistic quantum mechanics or relativistic quantum field theory—the latter theory being a quantum mechanical extension of relativistic continuum mechanics—would have to be considered. The condition that allows quantum effects to be safely ignored is that the sizes and separations of the bodies concerned are larger than their Compton wavelengths. (The Compton wavelength of a body of mass m is given by h/mc, where h is Planck's constant.) Despite these restrictions, there are nevertheless a number of situations in nature where relativistic mechanics is applicable. For example, it is essential to take into account the effects of relativity when calculating the motion of elementary particles accelerated to higher energies in particle accelerators, such as those at CERN (European Organization for Nuclear Research) near Geneva or at Fermilab (Fermi National Accelerator Laboratory) near Chicago. Moreover, such particles are caused to collide, thus creating further particles; although this creation process can only be understood through quantum mechanics, once the particles are well separated, they are subject to the laws of special relativity.

Similar remarks apply to cosmic rays that reach Earth from outer space. In some cases, these have energies as high as 10^{20} electron volts (eV). An electron of that energy has a velocity that differs from that of light by about 1 part in 10^{28}, as can be seen from the relativistic relation between energy and velocity, which will be given later. For a proton of the same energy, the velocity would differ from that of light by about 1 part in 10^{22}. At a more mundane level, relativistic mechanics must be used to calculate the energies of electrons or positrons emitted by the decay of radioactive nuclei. Astrophysicists need to use relativistic mechanics when dealing with the energy sources of stars, the energy released in supernova explosions, and the motion of electrons moving in the atmospheres of pulsars or when considering the hot big bang. At temperatures in the very early universe above 10^{10}kelvins (K), at which typical thermal energies kT (where k is Boltzmann's constant and T is temperature) are comparable with the rest mass energy of the electron, the primordial plasma must have been relativistic. Relativistic mechanics also must be considered when dealing with satellite navigational systems used, for example, by the military, such as the Global Positioning System (GPS). In this case, however, it is the purely kinematic effect on the rate of clocks on board the satellites (i.e., time dilation) that is important rather than the dynamic effects of relativity on the motion of the satellites themselves.

Relativistic Dynamics

Relativistic Energy and Momentum

There are a couple of (equivalent) ways to define momentum and energy in SR. One method uses conservation laws. If these laws are to remain valid in SR they must be true in every possible reference frame. However, if one does some simple thought experiments using the Newtonian definitions of momentum and energy, one sees that these quantities are not conserved in SR. One can rescue the idea of conservation by making some small modifications to the definitions to account for relativistic velocities. It is these new definitions which are taken as the correct ones for momentum and energy in SR.

The four-momentum of an object is straightforward, identical in form to the classical momentum, but replacing 3-vectors with 4-vectors:

$$\mathrm{P} = m_0 \mathrm{U} = (E / c, \mathrm{p})$$

The energy and momentum of an object with invariant mass m_0 (also called *rest mass*), moving with velocity v with respect to a given frame of reference, are respectively given by,

$$E = \gamma(\mathrm{v}) m_0 c^2$$
$$\mathrm{p} = \gamma(\mathrm{v}) m_0 \mathrm{v}$$

The factor of γ(v) comes from the definition of the four-velocity. The appearance of the γ factor has an alternative way of being stated.

The kinetic energy, *K*, is defined as,

$$K = (\gamma - 1) m_0 c^2 = E - m_0 c^2,$$

And the speed as a function of kinetic energy is given by,

$$v = \frac{c\sqrt{K(K + 2m_0 c^2)}}{K + m_0 c^2} = \frac{c\sqrt{(E - m_0 c^2)(E + m_0 c^2)}}{E} = \frac{pc^2}{E}.$$

Rest Mass and Relativistic Mass

The quantity,

$$m = \gamma(\mathrm{v})m_0$$

is often called the *relativistic mass* of the object in the given frame of reference.

This makes the relativistic relation between the spatial velocity and the spatial momentum look identical. However, this can be misleading, as it is *not* appropriate in special relativity in *all* circumstances. For instance, kinetic energy and force in special relativity can not be written exactly like their classical analogues by only replacing the mass with the relativistic mass. Moreover, under Lorentz transformations, this relativistic mass is not invariant, while the rest mass is. For this reason many people find it easier use the rest mass (thereby introduce γ through the 4-velocity or coordinate time), and discard the concept of relativistic mass.

Lev B. Okun suggested that “this terminology ... has no rational justification today”, and should no longer be taught.

Other physicists, including Wolfgang Rindler and T. R. Sandin, have argued that relativistic mass is a useful concept and there is little reason to stop using it.

Some authors use m for relativistic mass and m_0 for rest mass, others simply use m for rest mass.

The energy and momentum of an object with invariant mass m_0 are related by the formulas,

$$E^2 - (pc)^2 = (m_0 c^2)^2$$
$$\mathrm{p}c^2 = E\mathrm{v}.$$

The first is referred to as the *relativistic energy–momentum relation*. It can be derived by considering that $\frac{v^2}{c^2}$ can be written as $\frac{p^2}{\gamma^2 m_0^2 c^2}$ where the denominator can be written as $\frac{E^2}{c^2}$. Now, gamma can be replaced in the expression of energy. While the energy E and the momentum p depend on the frame of reference in which they are measured, the quantity $E^2 - (pc)^2$ is invariant, and arises as $-c^2$ times the squared magnitude of the 4-momentum vector which is $-(m_0 c)^2$.

It should be noted that the invariant mass of a system,

$$m_{\text{otot}} = \frac{\sqrt{E_{\text{tot}}^2 - (p_{\text{tot}} c)^2}}{c^2}$$

is different from the sum of the rest masses of the particles of which it is composed due to kinetic energy and binding energy. Rest mass is not a conserved quantity in special relativity unlike the situation in Newtonian physics. However, even if an object is changing internally, so long as it does not exchange energy with surroundings, then its rest mass will not change, and can be calculated with the same result in any frame of reference.

A particle whose rest mass is zero is called *massless*. Photons and gravitons are thought to be massless; and neutrinos are nearly so.

Mass–energy Equivalence

The relativistic energy–momentum equation holds for all particles, even for massless particles for which $m_0 = 0$. In this case:

$$E = pc$$

When substituted into $Ev = c^2p$, this gives $v = c$: massless particles (such as photons) always travel at the speed of light.

Notice that the rest mass of a composite system will generally be slightly different from the sum of the rest masses of its parts since, in its rest frame, their kinetic energy will increase its mass and their (negative) binding energy will decrease its mass. In particular, a hypothetical "box of light" would have rest mass even though made of particles which do not since their momenta would cancel.

Looking at the above formula for invariant mass of a system, one sees that, when a single massive object is at rest (v = 0, p = 0), there is a non-zero mass remaining: $m_0 = E/c^2$. The corresponding energy, which is also the total energy when a single particle is at rest, is referred to as "rest energy". In systems of particles which are seen from a moving inertial frame, total energy increases and so does momentum. However, for single particles the rest mass remains constant, and for systems of particles the invariant mass remain constant, because in both cases, the energy and momentum increases subtract from each other, and cancel. Thus, the invariant mass of systems of particles is a calculated constant for all observers, as is the rest mass of single particles.

The Mass of Systems and Conservation of Invariant Mass

For systems of particles, the energy–momentum equation requires summing the momentum vectors of the particles:

$$E^2 - \mathrm{p} \cdot \mathrm{p}c^2 = m_0^2 c^4$$

The inertial frame in which the momenta of all particles sums to zero is called the center of momentum frame. In this special frame, the relativistic energy–momentum equation has p= 0, and thus gives the invariant mass of the system as merely the total energy of all parts of the system, divided by c^2,

$$m_{0,\text{system}} = \sum_n E_n / c^2$$

This is the invariant mass of any system which is measured in a frame where it has zero total momentum, such as a bottle of hot gas on a scale. In such a system, the mass which the scale weighs is the invariant mass, and it depends on the total energy of the system. It is thus more than the sum of the rest masses of the molecules, but also includes all the totaled energies in the system as well. Like energy and momentum, the invariant mass of isolated systems cannot be changed so long as the system remains totally closed (no mass or energy allowed in or out), because the total relativistic energy of the system remains constant so long as nothing can enter or leave it.

An increase in the energy of such a system which is caused by translating the system to an inertial frame which is not the center of momentum frame, causes an increase in energy and momentum without an increase in invariant mass. $E = m_0c^2$, however, applies only to isolated systems in their center-of-momentum frame where momentum sums to zero.

Taking this formula at face value, we see that in relativity, mass is simply energy by another name (and measured in different units). In 1927 Einstein remarked about special relativity, "Under this theory mass is not an unalterable magnitude, but a magnitude dependent on (and, indeed, identical with) the amount of energy."

Closed (Isolated) Systems

In a "totally-closed" system (i.e., isolated system) the total energy, the total momentum, and hence the total invariant mass are conserved. Einstein's formula for change in mass translates to its simplest $\Delta E = \Delta mc^2$ form, however, only in non-closed systems in which energy is allowed to escape (for example, as heat and light), and thus invariant mass is reduced. Einstein's equation shows that such systems must lose mass, in accordance with the above formula, in proportion to the energy they lose to the surroundings. Conversely, if one can measure the differences in mass between a system before it undergoes a reaction which releases heat and light, and the system after the reaction when heat and light have escaped, one can estimate the amount of energy which escapes the system.

Chemical and Nuclear Reactions

In both nuclear and chemical reactions, such energy represents the difference in binding energies of electrons in atoms (for chemistry) or between nucleons in nuclei (in atomic reactions). In both cases, the mass difference between reactants and (cooled) products measures the mass of heat and light which will escape the reaction, and thus (using the equation) give the equivalent energy of heat and light which may be emitted if the reaction proceeds.

In chemistry, the mass differences associated with the emitted energy are around 10^{-9} of the molecular mass. However, in nuclear reactions the energies are so large that they are associated with mass differences, which can be estimated in advance, if the products and reactants have been weighed (atoms can be weighed indirectly by using atomic masses, which are always the same for each nuclide). Thus, Einstein's formula becomes important when one has measured the masses of different atomic nuclei. By looking at the difference in masses, one can predict which nuclei have stored energy that can be released by certain nuclear reactions, providing important information which was useful in the development of nuclear energy and, consequently, the nuclear bomb. Historically, for example, Lise Meitner was able to use the mass differences in nuclei to estimate that there was enough energy available to make nuclear fission a favorable process. The implications of this special form of Einstein's formula have thus made it one of the most famous equations in all of science.

Center of Momentum Frame

The equation $E = m_0c^2$ applies only to isolated systems in their center of momentum frame. It has been popularly misunderstood to mean that mass may be *converted* to energy, after which the *mass* disappears. However, popular explanations of the equation as applied to systems include open (non-isolated) systems for which heat and light are allowed to escape, when they otherwise would have contributed to the mass (invariant mass) of the system.

Historically, confusion about mass being "converted" to energy has been aided by confusion between mass and "matter", where matter is defined as fermion particles. In such a definition, electromagnetic radiation and kinetic energy (or heat) are not considered "matter". In some situations, matter may indeed be converted to non-matter forms of energy, but in all these situations, the matter and non-matter forms of energy still retain their original mass.

For isolated systems (closed to all mass and energy exchange), mass never disappears in the center of momentum frame, because energy cannot disappear. Instead, this equation, in context, means only that when any energy is added to, or escapes from, a system in the center-of-momentum frame, the system will be measured as having gained or lost mass, in proportion to energy added or removed. Thus, in theory, if an atomic bomb were placed in a box strong enough to hold its blast, and detonated upon a scale, the mass of this closed system would not change, and the scale would not move. Only when a transparent "window" was opened in the super-strong plasma-filled box, and light and heat were allowed to escape in a beam, and the bomb components to cool, would the system lose the mass associated with the energy of the blast. In a 21 kiloton bomb, for example, about a gram of light and heat is created. If this heat and light were allowed to escape, the remains of the bomb would lose a gram of mass, as it cooled. In this thought-experiment, the light and heat carry away the gram of mass, and would therefore deposit this gram of mass in the objects that absorb them.

Angular Momentum

In relativistic mechanics, the time-varying mass moment,

$$\mathrm{N} = m(\mathrm{x} - t\mathrm{v})$$

and orbital 3-angular momentum,

$$\mathrm{L} = \mathrm{x} \times \mathrm{p}$$

of a point-like particle are combined into a four-dimensional bivector in terms of the 4-position X and the 4-momentum P of the particle:

$$\mathrm{M} = \mathrm{X} \wedge \mathrm{P}$$

where, $\wedge$ denotes the exterior product. This tensor is additive: the total angular momentum of a system is the sum of the angular momentum tensors for each constituent of the system. So, for an assembly of discrete particles one sums the angular momentum tensors over the particles, or integrates the density of angular momentum over the extent of a continuous mass distribution.

Each of the six components forms a conserved quantity when aggregated with the corresponding components for other objects and fields.

Force

In special relativity, Newton's second law does not hold in the form F = ma, but it does if it is expressed as,

$$F = \frac{dp}{dt}$$

where p = γ(v)m_0v is the momentum as defined above and m_0 is the invariant mass. Thus, the force is given by,

$$F = \gamma(v)^3 m_0 a_{\parallel} + \gamma(v) m_0 a_{\perp}$$

Derivation

Starting from,

$$F = m_0 \frac{d(\gamma(v)v)}{dt} = m_0 \left(\frac{d\gamma(v)}{dt} v + \gamma(v) \frac{dv}{dt} \right).$$

Carrying out the derivatives gives,

$$F = \frac{\gamma(v)^3 m_0}{c^2} (v \cdot a) v + \gamma(v) m_0 a.$$

If the acceleration is separated into the part parallel to the velocity ($a_{\parallel}$) and the part perpendicular to it ($a_{\perp}$), so that:

$$a = a_{\parallel} + a_{\perp}, \quad v \cdot a_{\perp} = 0, \quad v \cdot a = v \cdot a_{\parallel},$$

one gets,

$$F = \frac{\gamma(v)^3 m_0}{c^2} (v \cdot a_{\parallel}) v + \gamma(v) m_0 (a_{\perp} + a_{\parallel}).$$

By construction $a_{\parallel}$ and v are parallel, so (v·$a_{\parallel}$)v is a vector with magnitude $v^2 a_{\parallel}$ in the direction of v (and hence $a_{\parallel}$) which allows the replacement:

$$(v \cdot a_{\parallel}) v = v^2 a_{\parallel}$$

then,

$$F = \frac{\gamma(v)^3 m_0 v^2}{c^2} a_{\parallel} + \gamma(v) m_0 (a_{\parallel} + a_{\perp})$$

$$= \gamma(v)^3 m_0 \left(\frac{v^2}{c^2} + \frac{1}{\gamma(v)^2} \right) a_{\parallel} + \gamma(v) m_0 a_{\perp}$$

$$= \gamma(v)^3 m_0 \left(\frac{v^2}{c^2} + 1 - \frac{v^2}{c^2} \right) a_{\parallel} + \gamma(v) m_0 a_{\perp}$$

$$= \gamma(v)^3 m_0 a_{\parallel} + \gamma(v) m_0 a_{\perp}$$

Consequently, in some old texts, $\gamma(v)^3 m_0$ is referred to as the *longitudinal mass,* and $\gamma(v)m_0$ is referred to as the *transverse mass,* which is numerically the same as the relativistic mass.

If one inverts this to calculate acceleration from force, one gets,

$$a = \frac{1}{m_0\gamma(v)}\left(F - \frac{(v\cdot F)v}{c^2}\right).$$

The force described in this section is the classical 3-D force which is not a four-vector. This 3-D force is the appropriate concept of force since it is the force which obeys Newton's third law of motion. It should not be confused with the so-called four-force which is merely the 3-D force in the comoving frame of the object transformed as if it were a four-vector. However, the density of 3-D force (linear momentum transferred per unit four-volume) *is* a four-vector (density of weight +1) when combined with the negative of the density of power transferred.

Torque

The torque acting on a point-like particle is defined as the derivative of the angular momentum tensor given above with respect to proper time:

$$\Gamma = \frac{dM}{d\tau} = X \wedge F$$

or in tensor components:

$$\Gamma_{\alpha\beta} = X_\alpha F_\beta - X_\beta F_\alpha$$

where F is the 4d force acting on the particle at the event X. As with angular momentum, torque is additive, so for an extended object one sums or integrates over the distribution of mass.

Kinetic Energy

The *work-energy theorem* says the change in kinetic energy is equal to the work done on the body. In special relativity:

$$\Delta K = W = [\gamma_1 - \gamma_0]m_0c^2.$$

Derivation

$$\begin{aligned}
\ddot{A}K = W &= \int_{x_0}^{x_1} F\cdot dx \\
&= \int_{t_0}^{t_1} \frac{d}{dt}(\gamma m_0 v)\cdot v dt \\
&= \gamma m_0 v\cdot v\Big|_{t_0}^{t_1} - \int_{t_0}^{t_1} \gamma m_0 v\cdot\frac{dv}{dt}dt \\
&= \gamma m_0 v^2\Big|_{t_0}^{t_1} - m_0\int_{v_0}^{v_1}\gamma v dv \\
&= m_0\left(\gamma v^2\Big|_{t_0}^{t_1} - c^2\int_{v_0}^{v_1}\frac{2v/c^2}{2\sqrt{1-v^2/c^2}}dv\right)
\end{aligned}$$

$$= m_0 \left(\frac{v^2}{\sqrt{1 - v^2/c^2}} + c^2 \sqrt{1 - v^2/c^2} \right) \Bigg|_{t_0}^{t_1}$$

$$= \frac{m_0 c^2}{\sqrt{1 - v^2/c^2}} \Bigg|_{t_0}^{t_1}$$

$$= \gamma m_0 c^2 \Big|_{t_0}^{t_1}$$

$$= \gamma_1 m_0 c^2 - \gamma_0 m c^2.$$

If in the initial state the body was at rest, so $v_0 = 0$ and $\gamma_0(v_0) = 1$, and in the final state it has speed $v_1 = v$, setting $\gamma_1(v_1) = \gamma(v)$, the kinetic energy is then;

$$K = [\gamma(v) - 1] m_0 c^2,$$

a result that can be directly obtained by subtracting the rest energy m_0c^2 from the total relativistic energy $\gamma(v)m_0c^2$.

Newtonian Limit

The Lorentz factor $\gamma(v)$ can be expanded into a Taylor series or binomial series for $(v/c)^2 < 1$, obtaining:

$$\gamma = \frac{1}{\sqrt{1-(v/c)^2}} = \sum_{n=0}^{\infty} \left(\frac{v}{c}\right)^{2n} \prod_{k=1}^{n} \left(\frac{2k-1}{2k}\right) = 1 + \frac{1}{2}\left(\frac{v}{c}\right)^2 + \frac{3}{8}\left(\frac{v}{c}\right)^4 + \frac{5}{16}\left(\frac{v}{c}\right)^6 + \cdots$$

and consequently,

$$E - m_0 c^2 = \frac{1}{2} m_0 v^2 + \frac{3}{8}\frac{m_0 v^4}{c^2} + \frac{5}{16}\frac{m_0 v^6}{c^4} + \cdots;$$

$$\mathrm{p} = m_0 \mathrm{v} + \frac{1}{2}\frac{m_0 v^2 \mathrm{v}}{c^2} + \frac{3}{8}\frac{m_0 v^4 \mathrm{v}}{c^4} + \frac{5}{16}\frac{m_0 v^6 \mathrm{v}}{c^6} + \cdots.$$

For velocities much smaller than that of light, one can neglect the terms with c^2 and higher in the denominator. These formulas then reduce to the standard definitions of Newtonian kinetic energy and momentum. This is as it should be, for special relativity must agree with Newtonian mechanics at low velocities.

Theory of Relativity

The Theory of Relativity, proposed by the Jewish physicist Albert Einstein (1879-1955) in the early part of the 20th century, is one of the most significant scientific advances of our time. Although the concept of relativity was not introduced by Einstein, his major contribution was the recognition that the speed of light in a vacuum is constant and an absolute physical boundary for motion. This does not have a major impact on a person's day-to-day life since we travel at speeds much slower than light speed. For objects travelling near light speed, however, the theory of relativity states that objects will move slower and shorten in length from the point of view of an observer on Earth. Einstein also derived the famous equation, $E = mc^2$, which reveals the equivalence of mass and energy.

When Einstein applied his theory to gravitational fields, he derived the "curved space-time continuum" which depicts the dimensions of space and time as a two-dimensional surface where massive objects create valleys and dips in the surface. This aspect of relativity explained the phenomena of light bending around the sun, predicted black holes as well as the Cosmic Microwave Background Radiation (CMB) -- a discovery rendering fundamental anomalies in the classic Steady-State hypothesis. For his work on relativity, the photoelectric effect and blackbody radiation, Einstein received the Nobel Prize in 1921.

Theory of Relativity – The Basics

Physicists usually dichotomize the Theory of Relativity into two parts.

- The first is the Special Theory of Relativity, which essentially deals with the question of whether rest and motion are relative or absolute, and with the consequences of Einstein's conjecture that they are relative.
- The second is the General Theory of Relativity, which primarily applies to particles as they accelerate, particularly due to gravitation, and acts as a radical revision of Newton's theory, predicting important new results for fast-moving and/or very massive bodies. The General Theory of Relativity correctly reproduces all validated predictions of Newton's theory, but expands on our understanding of some of the key principles. Newtonian physics had previously hypothesised that gravity operated through empty space, but the theory lacked explanatory power as far as how the distance and mass of a given object could be transmitted through space. General relativity irons out this paradox, for it shows that objects continue to move in a straight line in space-time, but we observe the motion as acceleration because of the curved nature of space-time.

Einstein's theories of both special and general relativity have been confirmed to be accurate to a very high degree over recent years, and the data has been shown to corroborate many key predictions; the most famous being the solar eclipse of 1919 bearing testimony that the light of stars is indeed deflected by the sun as the light passes near the sun on its way to earth. The total solar eclipse allowed astronomers to -- for the first time -- analyse starlight near the edge of the sun, which had been previously inaccessible to observers due to the intense brightness of the sun. It also predicted the rate at which two neutron stars orbiting one another will move toward each other. When this phenomenon was first documented, general relativity proved itself accurate to better than a trillionth of a percent precision, thus making it one of the best confirmed principles in all of physics.

Applying the principle of general relativity to our cosmos reveals that it is not static. Edwin Hubble (1889-1953) demonstrated in 1928 that the Universe is expanding, showing beyond reasonable doubt that the Universe sprang into being a finite time ago. The most common contemporary interpretation of this expansion is that this began to exist from the moment of the Big Bang some 13.7 billion years ago. However this is not the only plausible cosmological model which exists in academia, and many creation physicists such as Russell Humphreys and John Hartnett have devised models operating with a biblical framework, which -- to date -- have withstood the test of criticism from the most vehement of opponents.

A Testament to Creation

Using the observed cosmic expansion conjunctively with the general theory of relativity, we can infer from the data that the further back into time one looks, the universe ought to diminish in size accordingly. However, this cannot be extrapolated indefinitely. The universe's expansion helps us to appreciate the direction in which time flows. This is referred to as the Cosmological arrow of time, and implies that the future is -- by definition -- the direction towards which the universe increases in size. The expansion of the universe also gives rise to the second law of thermodynamics, which states that the overall entropy (or disorder) in the Universe can only increase with time because the amount of energy available for work deteriorates with time. If the universe was eternal, therefore, the amount of usable energy available for work would have already been exhausted. Hence it follows that at one point the entropy value was at absolute 0 (most ordered state at the moment of creation) and the entropy has been increasing ever since -- that is, the universe at one point was fully "wound up" and has been winding down ever since. This has profound theological implications, for it shows that time itself is necessarily finite. If the universe were eternal, the thermal energy in the universe would have been evenly distributed throughout the cosmos, leaving each region of the cosmos at uniform temperature (at very close to absolute 0), rendering no further work possible.

The General Theory of Relativity demonstrates that time is linked, or related, to matter and space, and thus the dimensions of time, space, and matter constitute what we would call a continuum. They must come into being at precisely the same instant. Time itself cannot exist in the absence of matter and space. From this, we can infer that the uncaused first cause must exist outside of the four dimensions of space and time, and possess eternal, personal, and intelligent qualities in order to possess the capabilities of intentionally space, matter -- and indeed even time itself -- into being.

Moreover, the very physical nature of time and space also suggest a Creator, for infinity and eternity must necessarily exist from a logical perspective. The existence of time implies eternity (as time has a beginning and an end), and the existence of space implies infinity. The very concepts of infinity and eternity infer a Creator because they find their very state of being in God, who transcends both and simply is.

Special Relativity

Special relativity is a fundamental physics theory about space and time that was developed by Albert Einstein in 1905 as a modification of Newtonian physics. It was created to deal with some pressing theoretical and experimental issues in the physics of the time involving light and electrodynamics. The predictions of special relativity correspond closely to those of Newtonian physics at speeds which are low in comparison to that of light, but diverge rapidly for speeds which are a significant fraction of the speed of light. Special relativity has been experimentally tested on numerous occasions since its inception, and its predictions have been verified by those tests.

Albert Einstein during a lecture in Vienna

Einstein postulated that the speed of light is the same for all observers, irrespective of their motion relative to the light source. This was in total contradiction to classical mechanics, which had been accepted for centuries. Einstein's approach was based on thought experiments and calculations. In 1908, Hermann Minkowski reformulated the theory based on different postulates of a more geometrical nature. His approach depended on the existence of certain interrelations between space and time, which were considered completely separate in classical physics. This reformulation set the stage for further developments of physics.

Special relativity makes numerous predictions that are incompatible with Newtonian physics (and everyday intuition). The first such prediction described by Einstein is called the relativity of simultaneity, under which observers who are in motion with respect to each other may disagree on whether two events occurred at the same time or one occurred before the other. The other major predictions of special relativity are time dilation (under which a moving clock ticks more slowly than when it is at rest with respect the observer), length contraction (under which a moving rod may be found to be shorter than when it is at rest with respect to the observer), and the equivalence of mass and energy (written as $E=mc^2$). Special relativity predicts a non-linear velocity addition formula, which prevents speeds greater than that of light from being observed. Special relativity also explains why Maxwell's equations of electromagnetism are correct in any frame of reference, and how an electric field and a magnetic field are two aspects of the same thing.

Special relativity has received experimental support in many ways, and it has been proven far more accurate than Newtonian mechanics. The most famous experimental support is the Michelson-Morley experiment, the results of which (showing that the speed of light is a constant) was one factor that motivated the formulation of the theory of special relativity. Other significant tests are the Fizeau experiment (which was first done decades before special relativity was proposed), the detection of the transverse Doppler effect, and the Haefele-Keating experiment. Today, scientists are so comfortable with the idea that the speed of light is always the same that the meter is now defined as being the distance traveled by light in $1/299,792,458^{th}$ of a second. This means that the speed of light is now defined as being 299, 792, 458 m/s.

Reference Frames and Galilean Relativity: A Classical Prelude

A reference frame is simply a selection of what constitutes stationary objects. Once the velocity of a certain object is arbitrarily defined to be zero, the velocity of everything else in the universe can be measured relative to it. When a train is moving at a constant velocity past a platform, one may either say that the platform is at rest and the train is moving or that the train is at rest and the platform is moving past it. These two descriptions correspond to two different reference frames. They are respectively called the rest frame of the platform and the rest frame of the train (sometimes simply the platform frame and the train frame).

The question naturally arises, can different reference frames be physically differentiated? In other words, can one conduct some experiments to claim that "we are now in an absolutely stationary reference frame?" Aristotle thought that all objects tend to cease moving and become at rest if there were no forces acting on them. Galileo challenged this idea and argued that the concept of absolute motion was unreal. All motion was relative. An observer who couldn't refer to some isolated object (if, say, he was imprisoned inside a closed spaceship) could never distinguish whether according to some external observer he was at rest or moving with constant velocity. Any experiment he could conduct would give the same result in both cases. However, accelerated reference frames are experimentally distinguishable. For example, if an astronaut moving in free space saw that the tea in his tea-cup was slanted rather than horizontal, he would be able to infer that his spaceship was accelerated. Thus not all reference frames are equivalent, but people have a class of reference frames, all moving at uniform velocity with respect to each other, in all of which Newton's first law holds. These are called the inertial reference frames and are fundamental to both classical mechanics and SR. Galilean relativity thus states that the laws of physics can not depend on absolute velocity, they must stay the same in any inertial reference frame. Galilean relativity is thus a fundamental principle in classical physics.

Mathematically, it says that if one transforms all velocities to a different reference frame, the laws of physics must be unchanged. What is this transformation that must be applied to the velocities? Galileo gave the common-sense "formula" for adding velocities: If

1. Particle P is moving at velocity v with respect to reference frame A and
2. Reference frame A is moving at velocity u with respect to reference frame B, then
3. The velocity of P with respect to B is given by v + u.

The formula for transforming coordinates between different reference frames is called the Galilean transformation. The principle of Galilean relativity then demands that laws of physics be unchanged if the Galilean transformation is applied to them. Laws of classical mechanics, like Newton's second law, obey this principle because they have the same form after applying the transformation. As Newton's law involves the derivative of velocity, any constant velocity added in a Galilean transformation to a different reference frame contributes nothing (the derivative of a constant is zero). Addition of a time-varying velocity (corresponding to an accelerated reference frame) will however change the formula, since Galilean relativity only applies to non-accelerated inertial reference frames.

Time is the same in all reference frames because it is absolute in classical mechanics. All observers measure exactly the same intervals of time and there is such a thing as an absolutely correct clock.

Invariance of Length: The Euclidean Picture

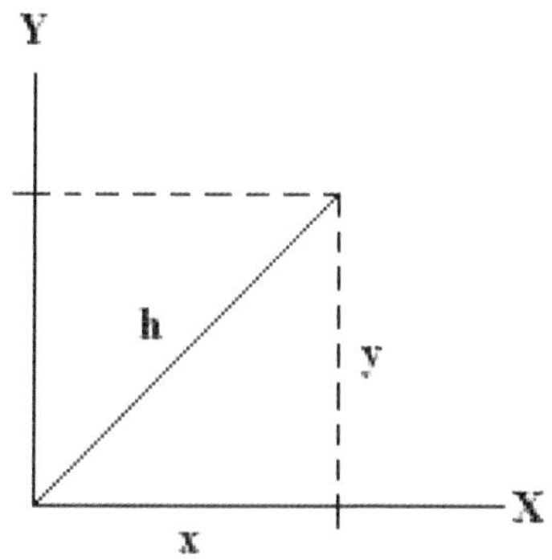

Pythagoras theorem

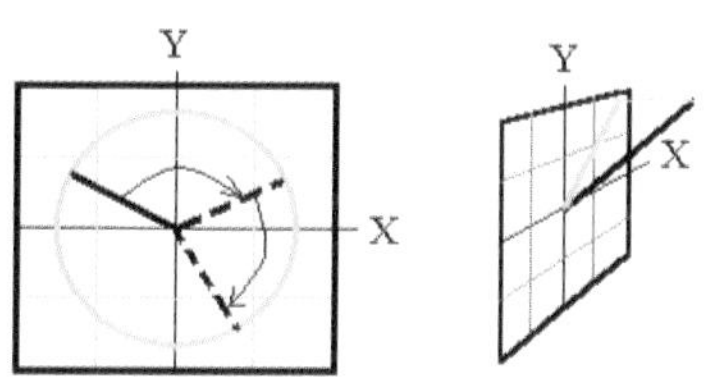

The length of an object is constant on the plane during rotations on the plane but not during rotations out of the plane

In special relativity, space and time are joined into a unified four-dimensional continuum called spacetime. To gain a sense of what spacetime is like, we must first look at the Euclidean space of Newtonian physics.

This approach to the theory of special relativity begins with the concept of "length." In everyday experience, it seems that the length of objects remains the same no matter how they are rotated or moved from place to place; as a result the simple length of an object doesn't appear to change or is "invariant." However, as is shown in the illustrations below, what is actually being suggested is that length seems to be invariant in a three-dimensional coordinate system.

The length of a line in a two-dimensional Cartesian coordinate system is given by Pythagoras' theorem:

$$h^2 = x^2 + y^2.$$

One of the basic theorems of vector algebra is that the length of a vector does not change when it is rotated. However, a closer inspection tells us that this is only true if we consider rotations confined to the plane. If we introduce rotation in the third dimension, then we can tilt the line out of the plane. In this case the projection of the line on the plane will get shorter. Does this mean length is not invariant? Obviously not. The world is three-dimensional and in a 3D Cartesian coordinate system the length is given by the three-dimensional version of Pythagoras's theorem:

$$k^2 = x^2 + y^2 + z^2.$$

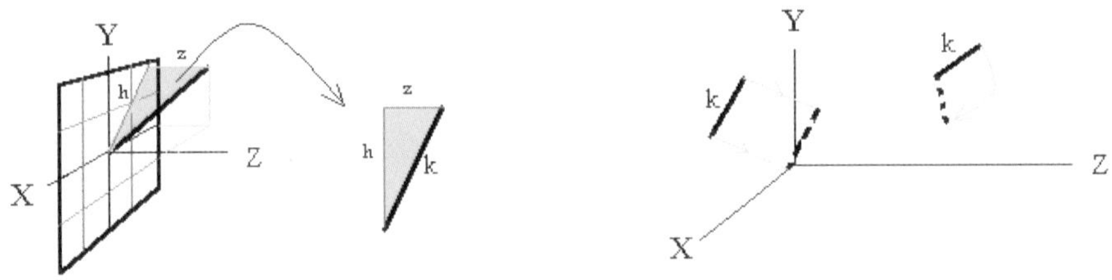

Invariance in a 3D coordinate system: Pythagoras theorem gives $k^2 = h^2 + z^2$ but $h^2 = x^2 + y^2$ therefore $k^2 = x^2 + y^2 + z^2$. The length of an object is constant whether it is rotated or moved from one place to another in a 3D coordinate system

This is invariant under all rotations. The apparent violation of invariance of length only happened because we were "missing" a dimension. It seems that, provided all the directions in which an object can be tilted or arranged are represented within a coordinate system, the length of an object does not change under rotations. A 3-dimensional coordinate system is enough in classical mechanics because time is assumed absolute and independent of space in that context. It can be considered separately.

Note that invariance of length is not ordinarily considered a dynamic principle, not even a theorem. It is simply a statement about the fundamental nature of space itself. Space as we ordinarily conceive it is called a three-dimensional Euclidean space, because its geometrical structure is described by the principles of Euclidean geometry. The formula for distance between two points is a fundamental property of an Euclidean space, it is called the Euclidean metric tensor (or simply the Euclidean metric). In general, distance formulas are called metric tensors.

Note that rotations are fundamentally related to the concept of length. In fact, one may define length or distance to be that which stays the same (is invariant) under rotations, or define rotations to be that which keep the length invariant. Given any one, it is possible to find the other. If we know the distance formula, we can find out the formula for transforming coordinates in a rotation. If, on the other hand, we have the formula for rotations then we can find out the distance formula.

Postulates of Special Relativity

Einstein developed Special Relativity on the basis of two postulates:

- First postulate—Special principle of relativity—The laws of physics are the same in all inertial frames of reference. In other words, there are no privileged inertial frames of reference.
- Second postulate—Invariance of c—The speed of light in a vacuum is independent of the motion of the light source.

Special Relativity can be derived from these postulates, as was done by Einstein in 1905. Einstein's postulates are still applicable in the modern theory but the origin of the postulates is more explicit. It was shown above how the existence of a universally constant velocity (the speed of light) is a consequence of modeling the universe as a particular four dimensional space having certain specific properties. The principle of relativity is a result of Minkowski structure being preserved under Lorentz transformations, which are *postulated* to be the physical transformations of inertial reference frames.

Minkowski Formulation: Introduction of Spacetime

Hermann Minkowski

After Einstein derived special relativity formally from the counterintuitive proposition that the speed of light is the same to all observers, the need was felt for a more satisfactory formulation. Minkowski, building on mathematical approaches used in non-Euclidean geometry and the mathematical work of Lorentz and Poincaré, realized that a geometric approach was the key. Minkowski showed in 1908 that Einstein's new theory could be explained in a natural way if the concept of separate space and time is replaced with one four-dimensional continuum called spacetime. This was a groundbreaking concept, and Roger Penrose has said that relativity was not truly complete until Minkowski reformulated Einstein's work.

The concept of a four-dimensional space is hard to visualize. It may help at the beginning to think simply in terms of coordinates. In three-dimensional space, one needs three real numbers to refer to a point. In the Minkowski space, one needs four real numbers (three space coordinates and one time coordinate) to refer to a point at a particular instant of time. This point at a particular instant of time, specified by the four coordinates, is called an event. The distance between two different events is called the spacetime interval.

A path through the four-dimensional spacetime, usually called Minkowski space, is called a world line. Since it specifies both position and time, a particle having a known world line has a completely determined trajectory and velocity. This is just like graphing the displacement of a particle moving in a straight line against the time elapsed. The curve contains the complete motional information of the particle.

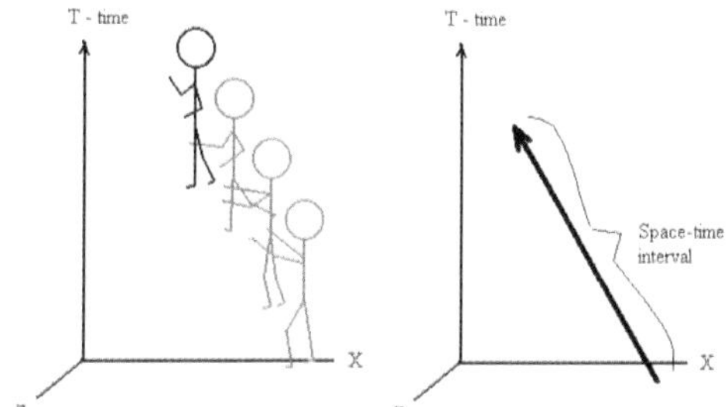

The spacetime interval

In the same way as the measurement of distance in 3D space needed all three coordinates we must include time as well as the three space coordinates when calculating the distance in Minkowski space (henceforth called M). In a sense, the spacetime interval provides a combined estimate of how far two events occur in space as well as the time that elapses between their occurrence.

But there is a problem. Time is related to the space coordinates, but they are not equivalent. Pythagoras's theorem treats all coordinates on an equal footing. We can exchange two space coordinates without changing the length, but we can not simply exchange a space coordinate with time, they are fundamentally different. It is an entirely different thing for two events to be separated in space and to be separated in time. Minkowski proposed that the formula for distance needed a change. He found that the correct formula was actually quite simple, differing only by a sign from Pythagoras's theorem:

$$s^2 = x^2 + y^2 + z^2 - (ct)^2$$

where c is a constant and t is the time coordinate. Multiplication by c, which has the dimension ms^{-1}, converts the time to units of length and this constant has the same value as the speed of light. So the spacetime interval between two distinct events is given by,

$$s^2 = (x_2 - x_1)^2 + (y_2 - y_1)^2 + (z_2 - z_1)^2 - c^2(t_2 - t_1)^2.$$

There are two major points to be noted. Firstly, time is being measured in the same units as length by multiplying it by a constant conversion factor. Secondly, and more importantly, the time-coordinate has a different sign than the space coordinates. This means that in the four-dimensional spacetime, one coordinate is different from the others and influences the distance differently. This new 'distance' may be zero or even negative. This new distance formula, called the metric of the spacetime, is at the heart of relativity. This distance formula is called the metric tensor of M. This minus sign means that a lot of our intuition about distances can not be directly carried over into spacetime intervals. For example, the spacetime interval between two events separated both in time and space may be zero. From now on, the terms distance formula and metric tensor will be used interchangeably, as will be the terms Minkowski metric and spacetime interval.

In Minkowski spacetime the spacetime interval is the invariant length, the ordinary 3D length is not required to be invariant. The spacetime interval must stay the same under rotations, but ordinary lengths can change. Just like before, we were missing a dimension. Note that everything this far are merely definitions. We define a four-dimensional mathematical construct which has a special formula for distance, where distance means that which stays the same under rotations (alternatively, one may define a rotation to be that which keeps the distance unchanged).

Now comes the physical part. Rotations in Minkowski space have a different interpretation than ordinary rotations. These rotations correspond to transformations of reference frames. Passing from one reference frame to another corresponds to rotating the Minkowski space. An intuitive justification for this is given below, but mathematically this is a dynamical postulate just like assuming that physical laws must stay the same under Galilean transformations (which seems so intuitive that we don't usually recognize it to be a postulate).

Since by definition rotations must keep the distance same, passing to a different reference frame must keep the spacetime interval between two events unchanged. This requirement can be used to derive an explicit mathematical form for the transformation that must be applied to the laws of physics (compare with the application of Galilean transformations to classical laws) when shifting reference frames. These transformations are called the Lorentz transformations. Just like the Galilean transformations are the mathematical statement of the principle of Galilean relativity in classical mechanics, the Lorentz transformations are the mathematical form of Einstein's principle of relativity. Laws of physics must stay the same under Lorentz transformations. Maxwell's equations and Dirac's equation satisfy this property, and hence, they are relativistically correct laws (but classically incorrect, since they don't transform correctly under Galilean transformations).

With the statement of the Minkowski metric, the common name for the distance formula given above, the theoretical foundation of special relativity is complete. The entire basis for special relativity can be summed up by the geometric statement "changes of reference frame correspond to rotations in the 4D Minkowski spacetime, which is defined to have the distance formula given above." The unique dynamical predictions of SR stem from this geometrical property of spacetime. Special relativity may be said to be the physics of Minkowski spacetime. In this case of spacetime, there are six independent rotations to be considered. Three of them are the standard rotations on a plane in two directions of space. The other three are rotations in a plane of both space and time: These rotations correspond to a change of velocity, and are described by the traditional Lorentz transformations.

One can replace distance formulas with rotation formulas. Instead of starting with the invariance of the Minkowski metric as the fundamental property of spacetime, one may state (as was done in classical physics with Galilean relativity) the mathematical form of the Lorentz transformations and require that physical laws be invariant under these transformations. This makes no reference to the geometry of spacetime, but will produce the same result. This was in fact the traditional approach to SR, used originally by Einstein himself. However, this approach is often considered to offer less insight and be more cumbersome than the more natural Minkowski formalism.

Reference Frames and Lorentz Transformations: Relativity Revisited

We have already discussed that in classical mechanics coordinate frame changes correspond to Galilean transfomations of the coordinates. Is this adequate in the relativistic Minkowski picture?

Suppose there are two people, Bill and John, on separate planets that are moving away from each other. Bill and John are on separate planets so they both think that they are stationary. John draws a graph of Bill's motion through space and time and this is shown in the illustration below.

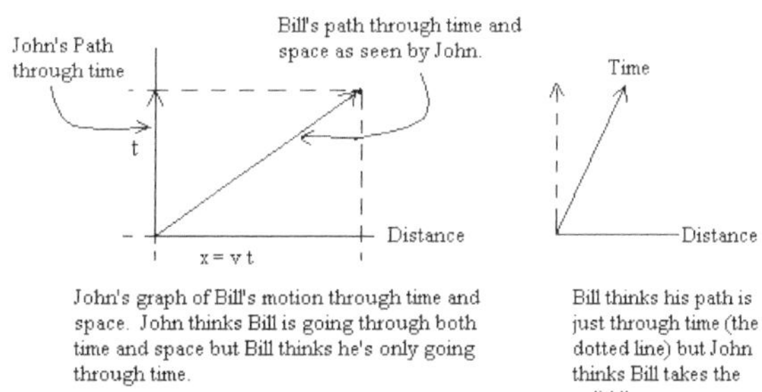

John's view of Bill and Bill's view of himself

John sees that Bill is moving through space as well as time but Bill thinks he is moving through time alone. Bill would draw the same conclusion about John's motion. In fact, these two views, which would be classically considered a difference in reference frames, are related simply by a coordinate transformation in M. Bill's view of his own world line and John's view of Bill's world line are related to each other simply by a rotation of coordinates. One can be transformed into the other by a rotation of the time axis. Minkowski geometry handles transformations of reference frames in a very natural way.

Changes in reference frame, represented by velocity transformations in classical mechanics, are represented by rotations in Minkowski space. These rotations are called Lorentz transformations. They are different from the Galilean transformations because of the unique form of the Minkowski metric. The Lorentz transformations are the relativistic equivalent of Galilean transformations. Laws of physics, in order to be relativistically correct, must stay the same under Lorentz transformations. The physical statement that they must be same in all inertial reference frames remains unchanged, but the mathematical transformation between different reference frames changes. Newton's laws of motion are invariant under Galilean rather than Lorentz transformations, so they are immediately recognizable as non-relativistic laws and must be discarded in relativistic physics. Schrödinger's equation is also non-relativistic.

Maxwell's equations are trickier. They are written using vectors and at first glance appear to transform correctly under Galilean transformations. But on closer inspection, several questions are apparent that can not be satisfactorily resolved within classical mechanics. They are indeed invariant under Lorentz transformations and are relativistic, even though they were formulated before the discovery of special relativity. Classical electrodynamics can be said to be the first relativistic theory in physics. To make the relativistic character of equations apparent, they are written using 4-component vector like quantities called 4-vectors. 4-Vectors transform correctly under Lorentz transformations. Equations written using 4-vectors are automatically relativistic. This is called the manifestly covariant form of equations. 4-Vectors form a very important part of the formalism of special relativity.

Einstein's Postulate: The Constancy of the Speed of Light

Einstein's postulate that the speed of light is a constant comes out as a natural consequence of the Minkowski formulation.

Proposition 1:

When an object is traveling at c in a certain reference frame, the spacetime interval is zero.

Proof:

The spacetime interval between the origin-event $(0, 0, 0, 0)$ and an event (x, y, z, t) is,

$$s^2 = x^2 + y^2 + z^2 - (ct)^2.$$

The distance travelled by an object moving at velocity v for t seconds is:

$$\sqrt{x^2 + y^2 + z^2} = vt$$

giving,

$$s^2 = (vt)^2 - (ct)^2.$$

Since the velocity v equals c we have

$$s^2 = (ct)^2 - (ct)^2.$$

Hence the spacetime interval between the events of departure and arrival is given by

$$s^2 = 0$$

Proposition 2:

An object traveling at c in one reference frame is traveling at c in all reference frames.

Proof:

Let the object move with velocity v when observed from a different reference frame. A change in reference frame corresponds to a rotation in M. Since the spacetime interval must be conserved under rotation, the spacetime interval must be the same in all reference frames. In proposition 1 we showed it to be zero in one reference frame, hence it must be zero in all other reference frames. We get that,

$$(vt)^2 - (ct)^2 = 0$$

which implies,

$$|v| = c.$$

The paths of light rays have a zero spacetime interval, and hence all observers will obtain the same value for the speed of light. Therefore, when assuming that the universe has four dimensions that are related by Minkowski's formula, the speed of light appears as a constant, and does not need to be assumed (postulated) to be constant as in Einstein's original approach to special relativity.

Clock Delays and Rod Contractions

Another consequence of the invariance of the spacetime interval is that clocks will appear to go slower on objects that are moving relative to you. This is very similar to how the 2D projection of a line rotated into the third-dimension appears to get shorter. Length is not conserved simply because we are ignoring one of the dimensions. Let us return to the example of John and Bill.

John observes the length of Bill's spacetime interval as:

$$s^2 = (vt)^2 - (ct)^2$$

whereas Bill doesn't think he has traveled in space, so writes:

$$s^2 = (0)^2 - (cT)^2$$

The spacetime interval, s^2, is invariant. It has the same value for all observers, no matter who measures it or how they are moving in a straight line. This means that Bill's spacetime interval equals John's observation of Bill's spacetime interval so:

$$(0)^2 - (cT)^2 = (vt)^2 - (ct)^2$$

and

$$-(cT)^2 = (vt)^2 - (ct)^2$$

hence

$$t = \frac{T}{\sqrt{1 - \frac{v^2}{c^2}}}.$$

So, if John sees a clock that is at rest in Bill's frame record one second, John will find that his own clock measures between these same ticks an interval *t,* called coordinate time, which is greater than one second. It is said that clocks in motion slow down, relative to those on observers at rest. This is known as "relativistic time dilation of a moving clock." The time that is measured in the rest frame of the clock (in Bill's frame) is called the proper time of the clock.

In special relativity, therefore, changes in reference frame affect time also. Time is no longer absolute. There is no universally correct clock, time runs at different rates for different observers.

Similarly it can be shown that John will also observe measuring rods at rest on Bill's planet to be shorter in the direction of motion than his own measuring rods. This is a prediction known as "relativistic length contraction of a moving rod." If the length of a rod at rest on Bill's planet is X, then we call this quantity the proper length of the rod. The length x of that same rod as measured on John's planet, is called coordinate length, and given by,

$$x = X\sqrt{1 - \frac{v^2}{c^2}}.$$

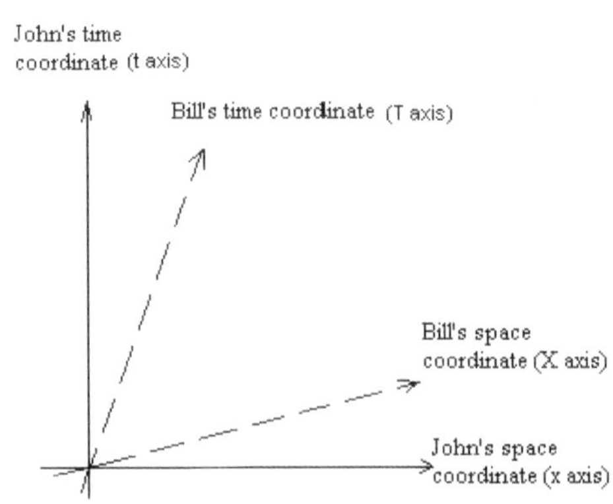

How Bill's coordinates appear to John at the instant they pass each other

These two equations can be combined to obtain the general form of the Lorentz transformation in one spatial dimension:

$$T = \gamma\left(t - \frac{vx}{c^2}\right)$$
$$X = \gamma(x - vt)$$

or equivalently:

$$t = \gamma\left(T + \frac{vX}{c^2}\right)$$
$$x = \gamma(X + vT)$$

where the Lorentz factor is given by $\gamma = \frac{1}{\sqrt{1 - v^2 / c^2}}$.

The above formulas for clock delays and length contractions are special cases of the general transformation.

Alternatively, these equations for time dilation and length contraction (here obtained from the invariance of the spacetime interval), can be *obtained* directly *from* the Lorentz transformation by setting X = 0 for time dilation, meaning that the clock is at rest in Bill's frame, or by setting t = 0 for length contraction, meaning that John must measure the distances to the end points of the moving rod at the same time.

A consequence of the Lorentz transformations is the modified velocity-addition formula:

$$s = \frac{v + u}{1 + (v / c)(u / c)}.$$

Simultaneity and Clock Desynchronization

Rather counter-intuitively, special relativity suggests that when 'at rest' we are actually moving through time at the speed of light. As we speed up in space we slow down in time. At the speed of light in space, time slows down to zero. This is a rotation of the time axis into the space axis. We observe that the object speeding by relativistically as having its time axis not at a right angle.

The consequence of this in Minkowski's spacetime is that clocks will appear to be out of phase with each other along the length of a moving object. This means that if one observer sets up a line of clocks that are all synchronized so they all read the same time, then another observer who is moving along the line at high speed will see the clocks all reading different times. This means that observers who are moving relative to each other see different events as simultaneous. This effect is known as "Relativistic Phase" or the "Relativity of Simultaneity." Relativistic phase is often overlooked by students of special relativity, but if it is understood, then phenomena such as the twin paradox are easier to understand.

Observers have a set of simultaneous events around them that they regard as composing the present instant. The relativity of simultaneity results in observers who are moving relative to each other having different sets of events in their present instant.

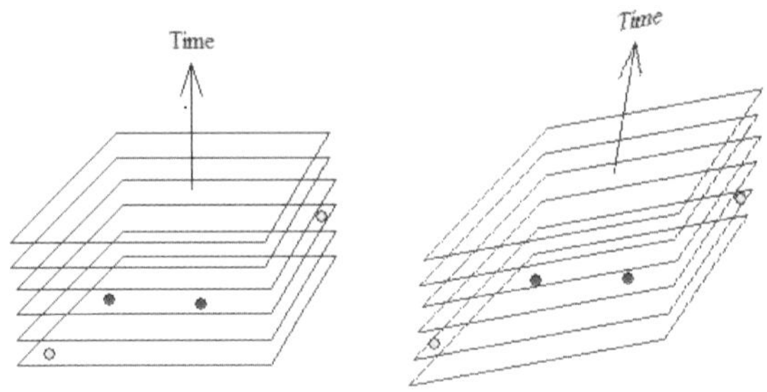

The "plane of simultaneity" or "surface of simultaneity" contains all those events that happen at the same instant for a given observer. Events that are simultaneous for one observer are not simultaneous for another observer in relative motion

The net effect of the four-dimensional universe is that observers who are in motion relative to you seem to have time coordinates that lean over in the direction of motion, and consider things to be simultaneous that are not simultaneous for you. Spatial lengths in the direction of travel are shortened, because they tip upwards and downwards, relative to the time axis in the direction of travel, akin to a rotation out of three-dimensional space.

Great care is needed when interpreting spacetime diagrams. Diagrams present data in two dimensions, and cannot show faithfully how, for instance, a zero length spacetime interval appears.

Mass Velocity Relationship

$E = mc^{22}$ where m stands for rest mass (invariant mass), applies most simply to single particles with no net momentum. But it also applies to ordinary objects composed of many particles so long as the particles are moving in different directions so the total momentum is zero. The mass of the object includes contributions from heat and sound, chemical binding energies and trapped radiation. Familiar examples are a tank of gas, or a hot bowl of soup. The kinetic energy of their particles, the heat motion and radiation, contribute to their weight on a scale according to $E = mc^{22}$.

The formula is the special case of the relativistic energy-momentum relationship:

$$E^2 - (pc)^2 = (mc^2)^2$$

This equation gives the rest mass of a system which has an arbitrary amount of momentum and energy. The interpretation of this equation is that the rest mass is the relativistic length of the energy-momentum four-vector.

If the equation $E = mc^2$ is used with the rest mass of the object, the E given by the equation will be the rest energy of the object, and will change with according to the object's internal energy, heat and sound and chemical binding energies, but will not change with the object's overall motion.

If the equation $E = mc^2$ is used with the relativistic mass of the object, the energy will be the total energy of the object, which is conserved in collisions with other fast moving objects.

In developing special relativity, Einstein found that the total energy of a moving body is,

$$E = \frac{m_0 c^2}{\sqrt{1 - \frac{v^2}{c^2}}},$$

with v the velocity.

For small velocities, this reduces to,

$$E = m_0 c^2 + \frac{1}{2} m_0 v^2 + \ldots$$

Which includes the newtonian kinetic energy, as expected, but also an enormous constant term, which is not zero when the object isn't moving.

The total momentum is:

$$P = \frac{m_0 v}{\sqrt{1 - \frac{v^2}{c^2}}}.$$

The ratio of the momentum to the velocity is the relativistic mass, and this ratio is equal to the total energy times c^2. The energy and relativistic mass are always related by the famous formula.

While this is suggestive, it does not immediately imply that the energy and mass are equivalent because the energy can always be redefined by adding or subtracting a constant. So it is possible to subtract the $m_0 c^2$ from the expression for E and this is also a valid conserved quantity, although an ugly one. Einstein needed to know whether the rest-mass of the object is really an energy, or whether the constant term was just a mathematical convenience with no physical meaning.

In order to see if the $m_0 c^2$ is physically significant, Einstein considered processes of emission and absorption. He needed to establish that an object loses mass when it emits energy. He did this by analyzing two photon emission in two different frames.

After Einstein first made his proposal, it became clear that the word mass can have two different meanings. The rest mass is what Einstein called m, but others defined the *relativistic mass* as:

$$m_{\text{rel}} = \frac{m_0}{\sqrt{1 - \frac{v^2}{c^2}}}.$$

This mass is the ratio of momentum to velocity, and it is also the relativistic energy divided by c^2. So the equation $E = m_{\text{rel}} c^2$ holds for moving objects. When the velocity is small, the relativistic mass and the rest mass are almost exactly the same.

$E = mc^2$ either means $E = m_0 c^2$ for an object at rest, or $E = m_{\text{rel}} c^2$ when the object is moving.

Einstein's original papers treated m as what would now be called the *rest mass* and some claim that he did not like the idea of "relativistic mass." When modern physicists say "mass, " they are usually talking about rest mass, since if they meant "relativistic mass, " they would just say "energy."

We can rewrite the expression for the energy as a Taylor series:

$$E = m_0c^2\left[1+\frac{1}{2}\left(\frac{v}{c}\right)^2+\frac{3}{8}\left(\frac{v}{c}\right)^4+\frac{5}{16}\left(\frac{v}{c}\right)^6+\ldots\right].$$

For speeds much smaller than the speed of light, higher-order terms in this expression get smaller and smaller because v/c is small. For low speeds we can ignore all but the first two terms:

$$E \approx m_0c^2+\frac{1}{2}m_0v^2.$$

The total energy is a sum of the rest energy and the Newtonian kinetic energy.

The classical energy equation ignores both the m_0c^2 part, and the high-speed corrections. This is appropriate, because all the high order corrections are small. Since only *changes* in energy affect the behavior of objects, whether we include the m_0c^2 part makes no difference, since it is constant. For the same reason, it is possible to subtract the rest energy from the total energy in relativity. In order to see if the rest energy has any physical meaning, it is essential to consider emission and absorption of energy in different frames.

The higher-order terms are extra correction to Newtonian mechanics which become important at higher speeds. The Newtonian equation is only a low speed approximation, but an extraordinarily good one. All of the calculations used in putting astronauts on the moon, for example, could have been done using Newton's equations without any of the higher order corrections.

Mass-energy Equivalence: Sunlight and Atom Bombs

Einstein showed that mass is simply another form of energy. The energy equivalent of rest mass m is $E = mc^2$. This equivalence implies that mass should be interconvertible with other forms of energy. This is the basic principle behind atom bombs and production of energy in nuclear reactors and stars (like Sun).

The standard model of the structure of matter has it that most of the 'mass' of the atom is in the atomic nucleus, and that most of this nuclear mass is in the intense field of light-like gluons swathing the quarks. Most of what is called the mass of an object is thus already in the form of energy, the energy of the quantum color field that confines the quarks.

The sun, for instance, fuels its prodigious output of energy by converting each second 600 billion kilograms of hydrogen-1 (single protons) into 595.2 billion kilograms of helium-4 (2 protons combined with 2 neutrons)—the 4.2 billion kilogram difference is the energy which the sun radiates into space each second. The sun, it is estimated, will continue to turn 4.2 billion kilos of mass into energy for the next 5 billion years or so before leaving the main sequence.

The atomic bombs that ended the Second World War, in comparison, converted about a thirtieth of an ounce of mass into energy.

The energy involved in chemical reactions is so small, however, that the conservation of mass is an excellent approximation.

General Relativity: A Peek Forward

Unlike Newton's laws of motion, relativity is not based upon dynamical postulates. It does not assume anything about motion or forces. Rather, it deals with the fundamental nature of spacetime. It is concerned with describing the geometry of the backdrop on which all dynamical phenomena take place. In a sense therefore, it is a meta-theory, a theory that lays out a structure that all other theories must follow. In truth, Special relativity is only a special case. It assumes that spacetime is flat. That is, it assumes that the structure of Minkowski space and the Minkowski metric tensor is constant throughout. In General relativity, Einstein showed that this is not true. The structure of spacetime is modified by the presence of matter. Specifically, the distance formula given above is no longer generally valid except in space free from mass. However, just like a curved surface can be considered flat in the infinitesimal limit of calculus, a curved spacetime can be considered flat at a small scale. This means that the Minkowski metric written in the differential form is generally valid.

$$ds^2 = dx^2 + dy^2 + dz^2 - c^2dt^2$$

One says that the Minkowski metric is valid locally, but it fails to give a measure of distance over extended distances. It is not valid globally. In fact, in general relativity the global metric itself becomes dependent on the mass distribution and varies through space. The central problem of general relativity is to solve the famous Einstein field equations for a given mass distribution and find the distance formula that applies in that particular case. Minkowski's spacetime formulation was the conceptual stepping stone to general relativity. His fundamentally new outlook allowed not only the development of general relativity, but also to some extent quantum field theories.

General Relativity

General relativity, part of the wide-ranging physical theory of relativity formed by the German-born physicist Albert Einstein. It was conceived by Einstein in 1916. General relativity is concerned with gravity, one of the fundamental forces in the universe. Gravity defines macroscopic behaviour, and so general relativity describes large-scale physical phenomena.

General relativity follows from Einstein's principle of equivalence: on a local scale it is impossible to distinguish between physical effects due to gravity and those due to acceleration. Gravity is treated as a geometric phenomenon that arises from the curvature of space-time. The solution of the fieldequations that describe general relativity can yield answers to different physical situations, such as planetary dynamics, the birth and death of stars, black holes, and the evolution of the universe. General relativity has been experimentally verified by observations of gravitational lenses, the orbit of the planet Mercury, the dilation of time in Earth's gravitational field, and gravitational waves from merging black holes.

German-Swiss-American physicist Albert Einstein showed that time and distance can be measured to have different values depending on their relative speed and direction. However, special relativity only applies to objects moving at a constant velocity. If an object accelerates, then the theory breaks down.

Einstein's theory of general relativity applies special relativity to the concept of acceleration.

English natural philosopher Isaac Newton had already shown the equivalence between gravitational acceleration and ordinary acceleration, and so general relativity unites theories of space and time with theories of light and matter.

The effects caused by acceleration and the effects caused by gravity are the same

Gravity can be equated with acceleration because the effects of a gravitational field can be produced by accelerating. In a closed room, there's no experiment a person can conduct that will tell them if objects are falling because they are experiencing a gravitational force, or because they are in deep space, accelerating at the same rate.

Einstein realised that a gravitational field would cause objects to accelerate if Minkowski's spacetime were curved. This acceleration is similar to when an object rolls down a curved surface, like a hill. In this case, however, the amount of acceleration is higher for objects with a higher mass because a higher mass causes a deeper, longer curve.

This means that objects accelerate because of the curvature of spacetime, and we call this effect gravity.

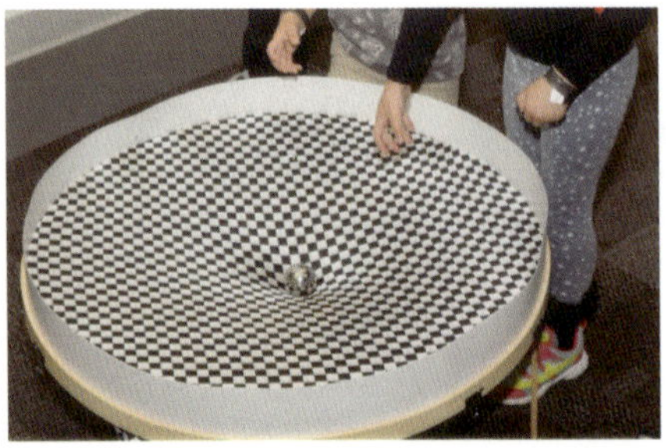

Massive objects, like stars, curve spacetime in a similar way to how massive objects, like a heavy ball, will curve a stretched-out piece of fabric. A lighter object, like a coin, may spin around the curve created by the ball, just as a lighter object like a planet may spin around the curve created by the Sun

The moon spins around the curve - or gravitational well - created by the Earth

The relationship between spacetime and mass was summarised by American physicist John Wheeler as follows:

> "Spacetime tells matter how to move; matter tells spacetime how to curve".

Einstein showed how the curvature of spacetime is related to the distribution of matter using equations developed by German mathematician Bernhard Riemann in the 1850s. Einstein showed that:

$$G_{\mu\nu} = R_{\mu\nu} - \frac{1}{2}Rg_{\mu\nu}$$

$$G_{\mu\nu} = \frac{8\pi G}{c^4}T_{\mu\nu}$$

Here, $G_{\mu\nu}$ is the Einstein tensor, which describes the curvature of spacetime. $R_{\mu\nu}$ is the Ricci tensor, which describes the relationship between Euclidean and non-Euclidean geometry. $g_{\mu\nu}$ is the metric tensor for Minkowski space, and describes all the intrinsic properties of the spacetime manifold, including time periods, distances, and volumes, as well as the curvature, and finally, $T_{\mu\nu}$ is the energy-momentum tensor, which describes the distribution of matter. The energy-momentum tensor is the source of the gravitational field, just as mass is the source of the gravitational field in Newton's equations.

A tensor is like a vector, which contains two properties. Velocity, for example, is a vector as it represents speed in a given direction. A tensor contains more than two properties, which may be written in a matrix - numbers or symbols that are arranged in rows and columns.

General relativity shows that observers in any frame will agree on how spacetime is curved by objects, and hence their gravitational field, whether they are moving relative the object or not. This means that the curvature of space, and hence the force of gravity is invariant.

Light moves through curved spacetime, taking the shortest possible path, however the shortest path across a curved surface is not necessarily a straight line. The shortest path, which may be curved, is known as a geodesic. This meant general relativity predicted that the path of light is bent by heavy objects, like the Sun.

The Cosmological Principle

In order to apply his theories to the universe as a whole, Einstein applied the cosmological principle. This assumes that the universe is homogeneous and isotropic when averaged over very large scales.

Homogeneity assumes that our observations are representational of the whole universe, and isotropy means that the universe is the same in whichever direction we look. These ideas were formulised by British astronomer Edward Milne in the 1930s, and verified by NASA's WMAP (Wilkinson Microwave Anisotropy Probe), which launched in 2001.

Confirmation of General Relativity

British physicist Arthur Eddington confirmed general relativity after the 1919 solar eclipse. Eddington knew that if mass curves spacetime, then light would travel in a curved path as it approaches a massive object like the Sun. This means that if the glare of the Sun were blocked out, like it is during an eclipse, he would be able to see stars that should be behind it. Einstein's theory predicted that the gravitational force of the Sun can cause starlight to deflect by up to 1.75 arc seconds (0.0005°), and Eddington confirmed this.

Eddington's photograph of the 1919 eclipse with dashed lines showing the positions of stars. The stars appeared to have moved from their usual relative positions because their light is bent around the gravitational well of the Sun

The light from distant galaxies is affected by the curvature of spacetime caused by a massive cluster of galaxies - Abell 2218. This has caused images of the galaxies to be enlarged and duplicated, and is known as gravitational lensing

Consequences of General Relativity

Precession of Mercury

The curvature of spacetime forces planets to orbit in open ellipses that are rotating. This effect is more noticeable the closer a planet is to the Sun, and so this explained why these effects were first observed in Mercury's orbit.

Gravitational Lensing

The curvature of spacetime means that the path of light is deflected around massive objects. This effect is known as gravitational lensing, and it can affect the shape of an event's light cone, allowing light to travel into previously forbidden regions. Gravitational lensing magnifies and brightens an image, and it's also possible for the same image to be projected more than once, as light is deflected in different directions. Eddington showed that light is deflected around the Sun in 1919.

Swiss astrophysicist Fritz Zwicky first considered using galaxy clusters as gravitational lenses in 1937.These effects were first observed by astronomers Dennis Walsh, Robert Carswell, and Ray Weymann, at the Kitt Peak National Observatory in the United States, in 1979.

Gravitational Waves

As massive objects move through spacetime, they will change its curvature, and this change will dissipate from the objects at the speed of light, creating 'ripples' in spacetime, like ripples on a pond. These ripples are known as gravitational waves. Scientists in the LIGO (Laser Interferometer Gravitational-Wave Observatory) Scientific Collaboration and the Virgo Collaboration discovered evidence for gravitational waves from a pair of merging black holes in 2016.

An artist's impression of a binary star system containing two neutron stars. The neutron stars create curves in spacetime as they orbit each other, and these dissipate at the speed of light, creating waves

Black Holes

Black holes occur when an object becomes so massive that even light cannot escape its gravitational well. We now know that supermassive black holes reside at the centre of most galaxies, including our own. These are millions of times the mass of the Sun.

The more massive the object, the more it curves spacetime, and the steeper the gravitational well. When an object is so massive that even light, which has no mass, cannot escape the well, the object is known as a 'black hole'

Time Travel

If mass curves space, then it will also curve time, and in 1949, Austrian mathematician Kurt Gödel suggested that time travel is possible because closed timelike curves can occur. These are regions where time is curved so that an observer would find themselves in an earlier time. Time travel would violate our notion of causality, and so there's still much debate over whether it's really possible.

Gravitational Time Dilation

By 1907, Einstein had already predicted that gravity would create an effect similar to the Doppler effect. Einstein predicted that from the point of view of a stationary observer, light will appear to be red shifted as it moves towards a strong gravitational force. This change corresponds to the fact that time appears to run slower for observers in a stronger gravitational field.

'Proper' time, or invariant time (t), is related to the relative time measured by something moving towards or away from a gravitational field (t_{rel}) via:

$$\Delta t_{rel} = \frac{\Delta t}{\sqrt{1 - \frac{2Gm}{rc^2}}}$$

Gravitational time dilation was first observed in 1971, when American physicists Joseph Hafele and Richard Keating flew four atomic clocks twice around the world on a commercial plane, and then compared their time to clocks left on Earth.

All satellite navigation systems, including Global Positioning Systems (GPS), have to take the effects of gravitational time dilation into account in order to be accurate.

Gravitational time dilation means that light appears redder as it moves towards a gravitational field

Absolute Space

Einstein suggested that general relativity disproves Newton's idea that space and time exist independently of matter and energy, as a type of substance, stating that general relativity "takes away the last remnants of physical objectivity from space and time".

This is because, in general relativity, spacetime is no longer considered a background in which events take place. Instead, spacetime is affected by what it contains. All observable events are described by the metric tensor, but there's no unique coordinate system. The choice is arbitrary. This means that the actual shape of spacetime can be described in a number of different ways that all give the same observable results. There's no experiment that can distinguish between them. If spacetime is a real substance, then it would presumably have to pick one of these shapes, and this leads to the 'hole' argument.

The hole argument shows that if you accept that coordinate systems do represent the true nature of spacetime, then there's no reason why there couldn't be a 'hole', where spacetime suddenly follows another coordinate system. There's no way to determine when a hole will appear, since both systems are observationally identical, and so proponents of absolute spacetime must accept that general relativity possesses the same objective uncertainty as quantum mechanics. If we remove the idea that spacetime is absolute, then we do not face this problem.

This does not necessarily address Newton's argument that absolute space can be derived from the fact that acceleration is absolute; we can always determine if we are accelerating. We cannot feel the effects of the movement of the Earth, for example, but we can prove that we are moving, and so some continue to argue that spacetime must be absolute.

A Non-static Universe

For Einstein, the most startling consequence of general relativity was that it shows that the shape of the universe depends on its energy density. This means that if all the mass in the universe is not sufficiently distributed, then everything will eventually fall in on itself, and the universe will end in a 'big crunch'. Einstein devised the cosmological constant in order to counterbalance this force and create a static universe.

In 1922, Russian mathematician Alexander Friedmann showed that it's also possible that the universe is expanding. This could occur if it begun at an extremely high density and temperature. If this were the case, and if mass is distributed too sparsely, then matter will continue to move apart forever.

American astronomer Edwin Hubble proved that the universe is expanding in 1929 . It's now thought that the universe will expand forever due to dark energy, which was discovered in 1998.

Advanced Concepts

Causal Structure and Global Geometry

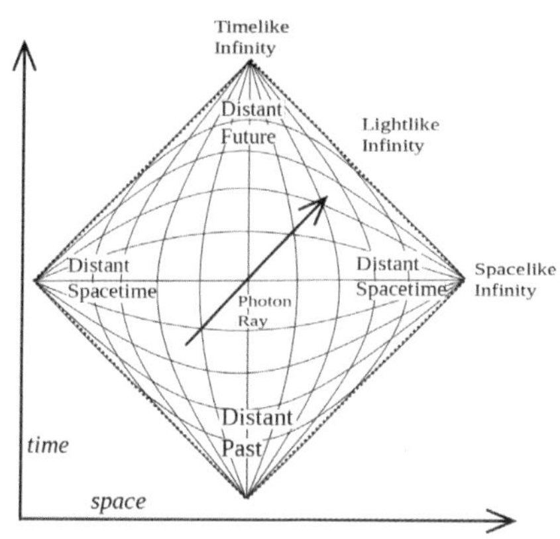

Penrose–Carter diagram of an infinite Minkowski universe

In general relativity, no material body can catch up with or overtake a light pulse. No influence from an event *A* can reach any other location *X* before light sent out at *A* to *X*. In consequence, an exploration of all light worldlines (null geodesics) yields key information about the spacetime's causal structure. This structure can be displayed using Penrose–Carter diagrams in which infinitely large regions of space and infinite time intervals are shrunk ("compactified") so as to fit onto a finite map, while light still travels along diagonals as in standard spacetime diagrams.

Aware of the importance of causal structure, Roger Penrose and others developed what is known as global geometry. In global geometry, the object of study is not one particular solution (or family of solutions) to Einstein's equations. Rather, relations that hold true for all geodesics, such as the Raychaudhuri equation, and additional non-specific assumptions about the nature of matter (usually in the form of energy conditions) are used to derive general results.

Horizons

Using global geometry, some spacetimes can be shown to contain boundaries called horizons, which demarcate one region from the rest of spacetime. The best-known examples are black holes: if mass is compressed into a sufficiently compact region of space (as specified in the hoop conjecture, the relevant length scale is the Schwarzschild radius), no light from inside can escape to the outside. Since no object can overtake a light pulse, all interior matter is imprisoned as well. Passage from the exterior to the interior is still possible, showing that the boundary, the black hole's *horizon*, is not a physical barrier.

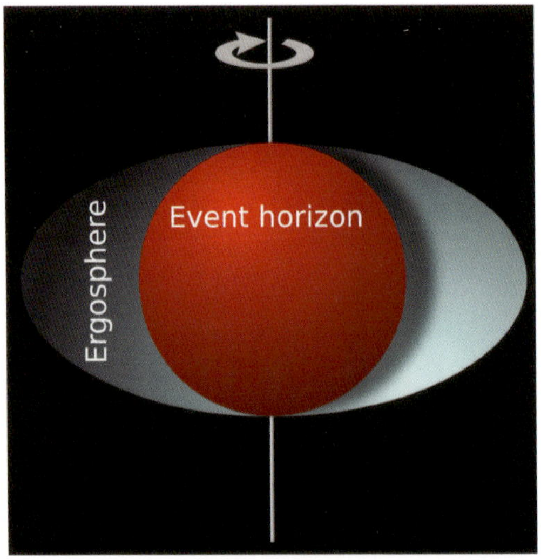

The ergosphere of a rotating black hole, which plays a key role when it comes to extracting energy from such a black hole

Early studies of black holes relied on explicit solutions of Einstein's equations, notably the spherically symmetric Schwarzschild solution (used to describe a static black hole) and the axisymmetric Kerr solution (used to describe a rotating, stationary black hole, and introducing interesting features such as the ergosphere). Using global geometry, later studies have revealed more general properties of black holes. In the long run, they are rather simple objects characterized by eleven parameters specifying energy, linear momentum, angular momentum, location at a specified time and electric charge. This is stated by the black hole uniqueness theorems: "black holes have no hair", that is, no distinguishing marks like the hairstyles of humans. Irrespective of the complexity of a gravitating object collapsing to form a black hole, the object that results (having emitted gravitational waves) is very simple.

Even more remarkably, there is a general set of laws known as black hole mechanics, which is analogous to the laws of thermodynamics. For instance, by the second law of black hole mechanics, the area of the event horizon of a general black hole will never decrease with time, analogous to the entropy of a thermodynamic system. This limits the energy that can be extracted by classical means from a rotating black hole (e.g. by the Penrose process). There is strong evidence that the laws of black hole mechanics are, in fact, a subset of the laws of thermodynamics, and that the black hole area is proportional to its entropy. This leads to a modification of the original laws of black hole mechanics: for instance, as the second law of black hole mechanics becomes part of the second law of thermodynamics, it is possible for black hole area to decrease—as long as other processes ensure that, overall, entropy increases. As thermodynamical objects with non-zero temperature, black holes should emit thermal radiation. Semi-classical calculations indicate that indeed they do, with the surface gravity playing the role of temperature in Planck's law. This radiation is known as Hawking radiation.

There are other types of horizons. In an expanding universe, an observer may find that some regions of the past cannot be observed ("particle horizon"), and some regions of the future cannot be influenced (event horizon). Even in flat Minkowski space, when described by an accelerated observer (Rindler space), there will be horizons associated with a semi-classical radiation known as Unruh radiation.

Singularities

Another general feature of general relativity is the appearance of spacetime boundaries known as singularities. Spacetime can be explored by following up on timelike and lightlike geodesics—all possible ways that light and particles in free fall can travel. But some solutions of Einstein's equations have "ragged edges"—regions known as spacetime singularities, where the paths of light and falling particles come to an abrupt end, and geometry becomes ill-defined. In the more interesting cases, these are "curvature singularities", where geometrical quantities characterizing spacetime curvature, such as the Ricci scalar, take on infinite values. Well-known examples of spacetimes with future singularities—where worldlines end—are the Schwarzschild solution, which describes a singularity inside an eternal static black hole, or the Kerr solution with its ring-shaped singularity inside an eternal rotating black hole. The Friedmann–Lemaître–Robertson–Walker solutions and other spacetimes describing universes have past singularities on which worldlines begin, namely Big Bang singularities, and some have future singularities (Big Crunch) as well.

Given that these examples are all highly symmetric—and thus simplified—it is tempting to conclude that the occurrence of singularities is an artifact of idealization. The famous singularity theorems, proved using the methods of global geometry, say otherwise: singularities are a generic feature of general relativity, and unavoidable once the collapse of an object with realistic matter properties has proceeded beyond a certain stage and also at the beginning of a wide class of expanding universes. However, the theorems say little about the properties of singularities, and much of current research is devoted to characterizing these entities' generic structure (hypothesized e.g. by the BKL conjecture).The cosmic censorship hypothesis states that all realistic future singularities (no perfect symmetries, matter with realistic properties) are safely hidden away behind a horizon, and thus invisible to all distant observers. While no formal proof yet exists, numerical simulations offer supporting evidence of its validity.

Evolution Equations

Each solution of Einstein's equation encompasses the whole history of a universe — it is not just some snapshot of how things are, but a whole, possibly matter-filled, spacetime. It describes the state of matter and geometry everywhere and at every moment in that particular universe. Due to its general covariance, Einstein's theory is not sufficient by itself to determine the time evolution of the metric tensor. It must be combined with a coordinate condition, which is analogous to gauge fixing in other field theories.

To understand Einstein's equations as partial differential equations, it is helpful to formulate them in a way that describes the evolution of the universe over time. This is done in "3+1" formulations, where spacetime is split into three space dimensions and one time dimension. The best-known example is the ADM formalism. These decompositions show that the spacetime evolution equations of general relativity are well-behaved: solutions always exist, and are uniquely defined, once suitable initial conditions have been specified.Such formulations of Einstein's field equations are the basis of numerical relativity.

Global and Quasi-local Quantities

The notion of evolution equations is intimately tied in with another aspect of general relativistic physics. In Einstein's theory, it turns out to be impossible to find a general definition for a seemingly simple property such as a system's total mass (or energy). The main reason is that the gravitational field—like any physical field—must be ascribed a certain energy, but that it proves to be fundamentally impossible to localize that energy.

Nevertheless, there are possibilities to define a system's total mass, either using a hypothetical "infinitely distant observer" (ADM mass) or suitable symmetries (Komar mass). If one excludes from the system's total mass the energy being carried away to infinity by gravitational waves, the result is the Bondi mass at null infinity. Just as in classical physics, it can be shown that these masses are positive. Corresponding global definitions exist for momentum and angular momentum. There have also been a number of attempts to define *quasi-local* quantities, such as the mass of an isolated system formulated using only quantities defined within a finite region of space containing that system. The hope is to obtain a quantity useful for general statements about isolated systems, such as a more precise formulation of the hoop conjecture.

Relationship with Quantum Theory

If general relativity were considered to be one of the two pillars of modern physics, then quantum theory, the basis of understanding matter from elementary particles to solid state physics, would be the other. However, how to reconcile quantum theory with general relativity is still an open question.

Quantum Field Theory in Curved Spacetime

Ordinary quantum field theories, which form the basis of modern elementary particle physics, are defined in flat Minkowski space, which is an excellent approximation when it comes to describing the behavior of microscopic particles in weak gravitational fields like those found on Earth. In order to describe situations in which gravity is strong enough to influence (quantum) matter, yet not strong enough to require quantization itself, physicists have formulated quantum field theories in curved spacetime. These theories rely on general relativity to describe a curved background spacetime, and define a generalized quantum field theory to describe the behavior of quantum matter within that spacetime.Using this formalism, it can be shown that black holes emit a blackbody spectrum of particles known as Hawking radiation leading to the possibility that they evaporate over time.As briefly mentioned above, this radiation plays an important role for the thermodynamics of black holes.

Quantum Gravity

The demand for consistency between a quantum description of matter and a geometric description of spacetime, as well as the appearance of singularities (where curvature length scales become microscopic), indicate the need for a full theory of quantum gravity: for an adequate description of the interior of black holes, and of the very early universe, a theory is required in which gravity and the associated geometry of spacetime are described in the language of quantum physics. Despite major efforts, no complete and consistent theory of quantum gravity is currently known, even though a number of promising candidates exist.

Projection of a Calabi–Yau manifold, one of the ways of compactifying the extra dimensions posited by string theory

Attempts to generalize ordinary quantum field theories, used in elementary particle physics to describe fundamental interactions, so as to include gravity have led to serious problems. Some have argued that at low energies, this approach proves successful, in that it results in an acceptable effective (quantum) field theory of gravity. At very high energies, however, the perturbative results are badly divergent and lead to models devoid of predictive power ("perturbative non-renormalizability").

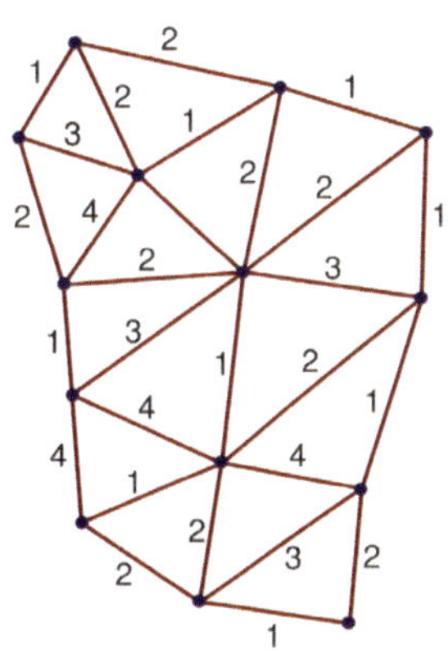

Simple spin network of the type used in loop quantum gravity

One attempt to overcome these limitations is string theory, a quantum theory not of point particles, but of minute one-dimensional extended objects. The theory promises to be a unified descriptionof all particles and interactions, including gravity; the price to pay is unusual features such as six extra dimensions of space in addition to the usual three. In what is called the second superstring revolution, it was conjectured that both string theory and a unification of general relativity and supersymmetry known as supergravity form part of a hypothesized eleven-dimensional model known as M-theory, which would constitute a uniquely defined and consistent theory of quantum gravity.

Another approach starts with the canonical quantization procedures of quantum theory. Using the initial-value-formulation of general relativity (cf. evolution equations above), the result is the Wheeler–deWitt equation (an analogue of the Schrödinger equation) which, regrettably, turns out to be ill-defined without a proper ultraviolet (lattice) cutoff. However, with the introduction of what are now known as Ashtekar variables, this leads to a promising model known as loop quantum gravity. Space is represented by a web-like structure called a spin network, evolving over time in discrete steps.

Depending on which features of general relativity and quantum theory are accepted unchanged, and on what level changes are introduced, there are numerous other attempts to arrive at a viable theory of quantum gravity, some examples being the lattice theory of gravity based on the Feynman Path Integral approach and Regge Calculus, dynamical triangulations, causal sets, twistor models or the path integral based models of quantum cosmology.

All candidate theories still have major formal and conceptual problems to overcome. They also face the common problem that, as yet, there is no way to put quantum gravity predictions to experimental tests (and thus to decide between the candidates where their predictions vary), although there is hope for this to change as future data from cosmological observations and particle physics experiments becomes available.

Current Status

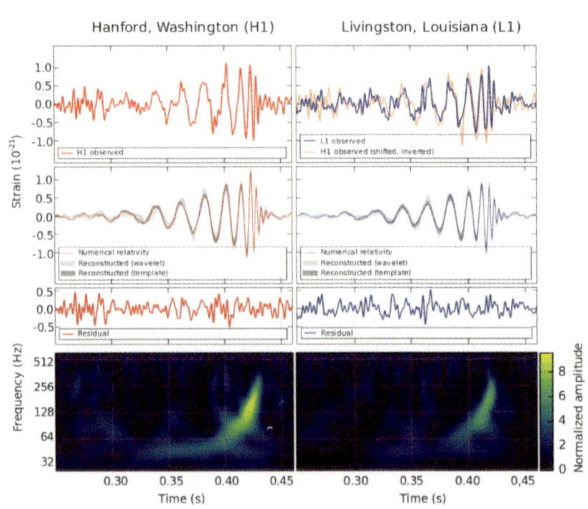

Observation of gravitational waves from binary black hole merger GW150914

General relativity has emerged as a highly successful model of gravitation and cosmology, which has so far passed many unambiguous observational and experimental tests. However, there are strong indications the theory is incomplete. The problem of quantum gravity and the question of the reality of spacetime singularities remain open. Observational data that is taken as evidence for dark energy and dark matter could indicate the need for new physics. Even taken as is, general relativity is rich with possibilities for further exploration. Mathematical relativists seek to understand the nature of singularities and the fundamental properties of Einstein's equations, while numerical relativists run increasingly powerful computer simulations (such as those describing merging black holes). In February 2016, it was announced that the existence of gravitational waves was directly detected by the Advanced LIGO team on September 14, 2015. A century after its introduction, general relativity remains a highly active area of research.

Relativistic Space-time

The concept of relativistic spacetime is fundamental to the structure of modern physics. The measurement of time is no longer absolute, but is, according to special relativity theory, path-dependent. In Einstein's terms, an observer at rest in one reference frame measures his own *proper time ("eigen-zeit")* on a clock, while observers moving relative to him can calculate his proper time using their measurements of both distance and time. From Albert Einstein's general relativity theory—the relativistic theory of gravitation—it is possible to derive three standard models of the Universe that allow for both finite and infinite temporal developments, all starting from an initial singularity. In deciding between these models, quantum mechanics—the modern (nonclassical) theory of matter—must be taken into account. Newton's conception of time now remains valid only locally and approximately for velocities considerably lower than the velocity of light.

Einstein's former mathematics professor, Hermann Minkowski, was perhaps the first to note this effect (and perhaps understood it even better than Einstein himself), and it was he who coined the phrase "space-time" to describe the interchangeability of the four dimensions. In 1908, Minkowski offered a useful analogy to help explain how four-dimensional space-time can appear differently to two observers in our normal three-dimensional space. He described two observers viewing a three-dimensional object from different angles, and noting that, for example, the length and width can appear different from the different viewpoints, due to what we call perspective, even though the object is clearly one and the same in three dimensions.

The idea perhaps becomes even clearer when we consider that our picture of the Moon is actually what the Moon was like 1¼ seconds ago (the time light takes to reach the Earth from the Moon), our picture of the Sun is actually how it looked 8½ minutes ago, and by the time we see an image of Alpha Centauri, our nearest star system, it is already 4.3 years out of date. We can therefore never know what the universe is like at this very instant, and the universe is clearly not a thing that extends just in space, but in space-time.

Due to the relativistic effects of time dilation, our idea of "now" is therefore something of a fictitious concept, one which we as humans have invented for ourselves, but for which nature itself has no real use. Physicists do not regard time as "passing" or "flowing" and time is not a sequence of events which happen: the past and the future are simply there, laid out as part of space-time.

But the malleability and blurring of space and time also has implications for other aspects of physics. Just as Maxwell had shown that the electric and magnetic fields, once considered completely separate entities, were both just part of a single seamless entity known as the electromagnetic field, likewise (although perhaps more difficult to grasp and perhaps more unexpected) energy and mass turn out to be just different faces of the same coin, a connection encapsulated in Einstein's justifiably famous formula, $E = mc^2$.

Relativistic Momentum

Momentum is one of the most important concepts in physics. The broadest form of Newton's second law is stated in terms of momentum. Momentum is conserved whenever the net external force on a system is zero. This makes momentum conservation a fundamental tool for analyzing collisions. All of Work, Energy, and Energy Resources is devoted to momentum, and momentum has been important for many other topics as well, particularly where collisions were involved. We will see that momentum has the same importance in modern physics. Relativistic momentum is conserved, and much of what we know about subatomic structure comes from the analysis of collisions of accelerator-produced relativistic particles.

The first postulate of relativity states that the laws of physics are the same in all inertial frames. Does the law of conservation of momentum survive this requirement at high velocities? The answer is yes, provided that the momentum is defined as follows.

Relativistic momentum is defined in such a way that the conservation of momentum will hold in all inertial frames. Whenever the net external force on a system is zero, relativistic momentum is conserved, just as is the case for classical momentum. This has been verified in numerous experiments.

In Relativistic Energy, the relationship of relativistic momentum to energy is explored. That subject will produce our first inkling that objects without mass may also have momentum.

Classically, momentum is defined by,

$$\vec{p} = m\vec{u}$$

where m and $\vec{u}$ the mass and velocity of an object. What is the correct relativistic definition of momentum? We have the freedom to define momentum any way we like, but to be useful, we would like to define it such that total momentum is conserved in all inertial frames.

Recall the law of momentum conservation. If there are n bodies with momenta $\vec{p}_1, \ldots, \vec{p}_n$, then, in the absence of external forces, the total momentum is given by,

$$\sum \vec{p} = \vec{p}_1 + \ldots + \vec{p}_n$$

cannot change. We would like to define momentum relativistically to preserve conservation of momentum.

We can't use the classical definition Eq. $\vec{p} = m\vec{u}$ because of the way velocity transforms between reference frames. For example, the way u_y transforms in going from S to S ′ depends on u_x. The classical definition of momentum is,

$$\vec{p} = m\vec{u} = m\frac{d\vec{r}}{dt}$$

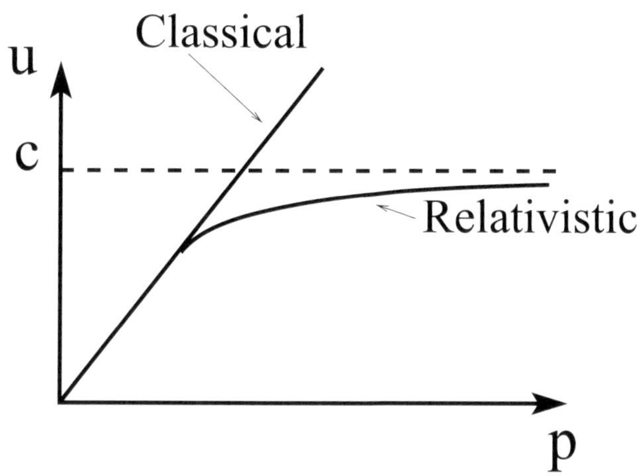

Velocity versus momentum

The problem is that both $\vec{r}$ and t are subject to the Lorentz transformation and that makes things messy.

The relativistic way to define momentum is,

$$\vec{p} = m\frac{d\vec{r}}{dt_o}$$

where t_o is the proper time, i.e., the time measured in the rest frame of the particle or object. Notice that with this definition,

$$p_y = m\frac{dy}{dt_o}$$

which is invariant when one goes between reference frames that are moving relative to one another along the x–axis. That is because t_o is invariant and y = y ′. Similarly for p_z.

Since dt = γdt_o where $\gamma = 1\sqrt{1-u^2/c^2}$, we can write,

$$\vec{p} = \frac{m\vec{u}}{\sqrt{1-u^2/c^2}} = \gamma m\vec{u}$$

A constant force changes the momentum and hence the velocity. Notice that as u → c, γ and hence p increase without limit, but the speed u never reaches the speed of light. Thus no object can go faster than the speed of light.

We can define a relativistic mass M(u) by,

$$M(u) = \frac{M}{\sqrt{1-u^2/c^2}} = M\gamma$$

calls this a variable mass m_{var}, but it doesn't want to use this. However, it does make it easy to write the momentum in the traditional form,

$$\vec{p} = M(u)\vec{u}$$

Notice from Eq.mentioned before the above one, that if the speed u of an object with nonzero mass is equal to the speed of light, the relativistic mass is infinity. If the object's speed exceeds the speed of light, then the relativistic mass M(u) is imaginary. Clearly, this is absurd and so, once again, the velocity of an object cannot exceed the speed of light. If the object has mass, it cannot travel at the speed of light; its speed must be less than the speed of light.

Relativistic Mass

Relativistic mass refers to mass of a body which change with the speed of the body as this speeds approaches close to speed of light, it increases with velocity and tends to infinity when the velocity approaches the speed of light.

Concept of Relativistic Mass

In special relativity, the expressions for the total energy and momentum of an object can be written as,

$$E = m_0\gamma c \;, \vec{p} = m_0\gamma\vec{v},$$

where $\gamma = \frac{1}{\sqrt{1-v^2/c^2}}$ and m_0 is the invariant mass (or rest mass) of the object. If we introduce the relativistic mass,

$$m = m_0\gamma$$

these expressions turn into,

which many find more familiar since the first is an iconic relation in special relativity and the latter coincides with the classical expression for momentum.

As almost everyone knows, Newton's 2nd law says that,

$$F = ma$$

where F is the force applied, m is the mass, and a is the acceleration felt by the body. In Newtonian mechanics, mass is invariant, but a consequence of special relativity is that nothing can travel faster than the speed of light c. This raises the conundrum of why can't we keep applying a force to a body of mass m, causing it to continue accelerating and to ultimately increase its velocity to one greater than the speed of light?

The answer is that Newton's 2nd law is incomplete. Einstein showed that mass is also a function of velocity, and so we should write,

$$m = \gamma m_0$$

Where $\gamma = \frac{1}{\sqrt{(1 - V^2 / c^2)}}$ is the so-called *Lorentz factor* and m_0 is the rest mass (also known as the invariant mass or gravitational mass), the mass an object has when it is at rest relative to the observer. Hence we can argue that, as we approach the speed of light, the applied force goes into changing the mass of the body, rather than accelerating it, leading to a modified version of Newton's 2nd law,

$$F \quad m \, a$$

where both velocity and/or mass change as a force is applied. But, because of the fact that $\gamma \approx 1$ until $V \approx c / 2$, very little increase in mass occurs until V has reached appreciable values.

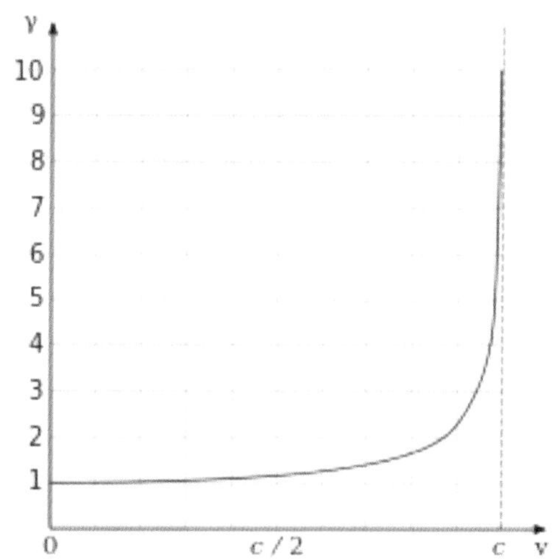

The variation of γ (the Lorentz factor) as a function of the speed V. Until $V \approx c / 2, \gamma$ is very close to unity

Relativistic Energy

The relativistic energy is the way that Einstein showed that the law of conservation of energy is valid relativistically, it means, the law of conservation of energy is valid in all inertial frames in high velocities approaching to the speed of light.

Relativistic energy = rest mass * speed of light squared / squared root [one minus (velocity / speed of light) squared].

The equation is:

$$E = mc^2 / \text{sqrt}\,(1 - v^2 / c^2)$$

Where:

E: relativistic energy

m: rest mass (invariant mass)

v: velocity of the body

c: speed of light

Conservation of energy is one of the most important laws in physics. Not only does energy have many important forms, but each form can be converted to any other. We know that classically the total amount of energy in a system remains constant. Relativistically, energy is still conserved, provided its definition is altered to include the possibility of mass changing to energy, as in the reactions that occur within a nuclear reactor. Relativistic energy is intentionally defined so that it will be conserved in all inertial frames, just as is the case for relativistic momentum. As a consequence, we learn that several fundamental quantities are related in ways not known in classical physics. All of these relationships are verified by experiment and have fundamental consequences. The altered definition of energy contains some of the most fundamental and spectacular new insights into nature found in recent history.

Total Energy and Rest Energy

The first postulate of relativity states that the laws of physics are the same in all inertial frames. Einstein showed that the law of conservation of energy is valid relativistically, if we define energy to include a relativistic factor.

Total Energy

Total energy E is defined to be $E = \gamma mc^2$, where m is mass, c is the speed of light,

$$\gamma = \frac{1}{\sqrt{1 - \frac{v^2}{c^2}}}$$

and v is the velocity of the mass relative to an observer. There are many aspects of the total energy E that we will discuss—among them are how kinetic and potential energies are included in E, and how E is related to relativistic momentum. But first, note that at rest, total energy is not zero. Rather, when $v = 0$, we have $\gamma = 1$, and an object has rest energy.

Rest Energy

Rest energy is $E_0 = mc^2$.

This is the correct form of Einstein's most famous equation, which for the first time showed that energy is related to the mass of an object at rest. For example, if energy is stored in the object, its rest mass increases. This also implies that mass can be destroyed to release energy. The implications of these first two equations regarding relativistic energy are so broad that they were not completely recognized for some years after Einstein published them in 1907, nor was the experimental proof that they are correct widely recognized at first. Einstein, it should be noted, did understand and describe the meanings and implications of his theory.

Kinetic Energy and the Ultimate Speed Limit

Kinetic energy is energy of motion. Classically, kinetic energy has the familiar expression $\frac{1}{2}mv^2$. The relativistic expression for kinetic energy is obtained from the work-energy theorem. This theorem states that the net work on a system goes into kinetic energy. If our system starts from rest, then the work-energy theorem is W_{net} = KE.

Relativistically, at rest we have rest energy $E_0 = mc^2$. The work increases this to the total energy $E = \gamma mc^2$. Thus, $W_{net} = E - E_0 = \gamma mc^2 - mc^2 = (\gamma - 1)mc^2$.

Relativistically, we have $W_{net} = KE_{rel}$.

Relativistic Kinetic Energy

Relativistic kinetic energy is $KE_{rel} = (\gamma - 1)mc^2$.

When motionless, we have $v = 0$ and

$$\gamma = \frac{1}{\sqrt{1 - \frac{v^2}{c^2}}} = 1$$

so that $KE_{rel} = 0$ at rest, as expected. But the expression for relativistic kinetic energy (such as total energy and rest energy) does not look much like the classical $\frac{1}{2}mv^2$. To show that the classical expression for kinetic energy is obtained at low velocities, we note that the binomial expansion for γ at low velocities gives,

$$\gamma = 1 + \frac{1v^2}{2c^2}$$

A binomial expansion is a way of expressing an algebraic quantity as a sum of an infinite series of terms. In some cases, as in the limit of small velocity here, most terms are very small. Thus the expression derived for γ here is not exact, but it is a very accurate approximation. Thus, at low velocities,

$$\gamma - 1 = \frac{1v^2}{2c^2}$$

Entering this into the expression for relativistic kinetic energy gives,

$$KE_{rel} = \left(\frac{1v^2}{2c^2}\right)mc^2 = \frac{1}{2}mv^2 = KE_{class}$$

So, in fact, relativistic kinetic energy does become the same as classical kinetic energy when $v << c$.

It is even more interesting to investigate what happens to kinetic energy when the velocity of an object approaches the speed of light. We know that γ becomes infinite as v approaches c, so that KE_{rel} also becomes infinite as the velocity approaches the speed of light. An infinite amount of work (and, hence, an infinite amount of energy input) is required to accelerate a mass to the speed of light.

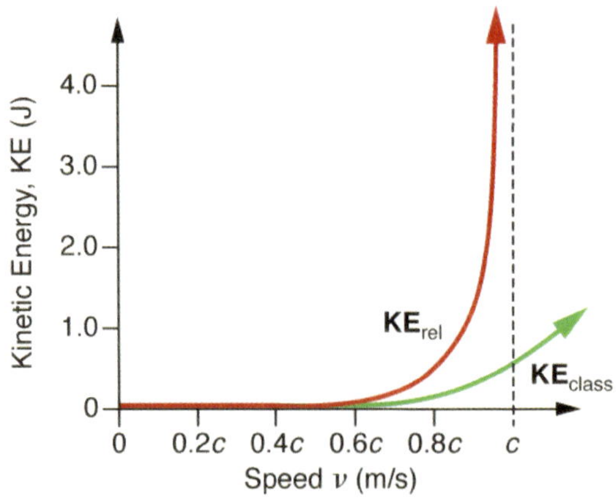

This graph of KE_{rel} versus velocity shows how kinetic energy approaches infinity as velocity approaches the speed of light. It is thus not possible for an object having mass to reach the speed of light. Also shown is KE_{class}, the classical kinetic energy, which is similar to relativistic kinetic energy at low velocities. Note that much more energy is required to reach high velocities than predicted classically

So the speed of light is the ultimate speed limit for any particle having mass. All of this is consistent with the fact that velocities less than c always add to less than c. Both the relativistic form for kinetic energy and the ultimate speed limit being c have been confirmed in detail in numerous experiments. No matter how much energy is put into accelerating a mass, its velocity can only approach—not reach—the speed of light.

Example

Comparing Kinetic Energy: Relativistic Energy Versus Classical Kinetic Energy

An electron has a velocity $v = 0.990c$.

1. Calculate the kinetic energy in MeV of the electron.
2. Compare this with the classical value for kinetic energy at this velocity. (The mass of an electron is 9.11×10^{-31} kg.)

Strategy

The expression for relativistic kinetic energy is always correct, but for Part 1 it must be used since the velocity is highly relativistic (close to c). First, we will calculate the relativistic factor γ, and then use it to determine the relativistic kinetic energy. For Part 2, we will calculate the classical kinetic energy (which would be close to the relativistic value if v were less than a few percent of c) and see that it is not the same.

Solution for Part 1:

Identify the knowns: $v = 0.990c$; $m = 9.11 \times 10^{-31}$ kg

Identify the unknown: KE_{rel}

Choose the appropriate equation. $KE_{rel} = (\gamma - 1)mc^2$

Plug the knowns into the equation.First calculate γ. We will carry extra digits because this is an intermediate calculation.

$$\begin{aligned}\gamma &= \frac{1}{\sqrt{1-\frac{v^2}{c^2}}} \\ &= \frac{1}{\sqrt{1-\frac{(0.990c)^2}{c^2}}} \\ &= \frac{1}{\sqrt{1-(0.990)^2}} \\ &= 7.0888\end{aligned}$$

Next, we use this value to calculate the kinetic energy.

$$\begin{aligned}KE_{rel} &= (\gamma - 1)mc^2 \\ &= (7.0888-1)(9.11\times10^{-31}\,\text{kg})(3.00\times10^8\,m/s)^2 \\ &= 4.99\times10^{13}\,\text{J}\end{aligned}$$

Convert units.

$$\begin{aligned}KE_{rel} &= (4.99\times10^{13}\,\text{J})\left(\frac{1\,\text{MeV}}{1.60\times10^{-13}\,\text{J}}\right) \\ &= 3.12\text{MeV}\end{aligned}$$

Solution for Part 2:

List the knowns: $v = 0.990c$; $m = 9.11 \times 10^{-31}$ kg

List the unknown: KE_{class}

Choose the appropriate equation.

$$KE_{class} = \frac{1}{2}mv^2$$

Plug the knowns into the equation.

$$KE_{class} = \frac{1}{2}mv^2$$

$$= \frac{1}{2}(9.00\times10^{-31}\text{kg})(0.990)^2\left(3.00\times10^{8}m/s\right)^2$$
$$= 4.02\times10^{-14}\text{J}$$

Convert units.

$$\text{KE}_{\text{class}} = 4.02\times10^{-14}\text{J}\left(\frac{1\text{ MeV}}{1.60\times10^{-13}\text{J}}\right)$$
$$= 0.251\text{MeV}$$

As might be expected, since the velocity is 99.0% of the speed of light, the classical kinetic energy is significantly off from the correct relativistic value. Note also that the classical value is much smaller than the relativistic value. In fact,

$$\frac{\text{KE}_{\text{rel}}}{\text{KE}_{\text{class}}} = 12.4$$

here. This is some indication of how difficult it is to get a mass moving close to the speed of light. Much more energy is required than predicted classically.

The Fermi National Accelerator Laboratory, near Batavia, Illinois, was a subatomic particle collider that accelerated protons and antiprotons to attain energies up to 1 Tev (a trillion electronvolts). The circular ponds near the rings were built to dissipate waste heat

Some people interpret this extra energy as going into increasing the mass of the system, but, as discussed in Relativistic Momentum, this cannot be verified unambiguously. What is certain is that ever-increasing amounts of energy are needed to get the velocity of a mass a little closer to that of light. An energy of 3 MeV is a very small amount for an electron, and it can be achieved with present-day particle accelerators. SLAC, for example, can accelerate electrons to over 50×10^9 eV = 50, 000 MeV.

Is there any point in getting v a little closer to c than 99.0% or 99.9%? The answer is yes. We learn a great deal by doing this. The energy that goes into a high-velocity mass can be converted to any other form, including into entirely new masses. Most of what we know about the substructure of matter and the collection of exotic short-lived particles in nature has been learned this way.

Particles are accelerated to extremely relativistic energies and made to collide with other particles, producing totally new species of particles. Patterns in the characteristics of these previously unknown particles hint at a basic substructure for all matter.

Relativistic Energy and Momentum

We know classically that kinetic energy and momentum are related to each other, since,

$$KE_{class} = \frac{p^2}{2m} = \frac{(mv)^2}{2m} = \frac{1}{2}mv^2$$

Relativistically, we can obtain a relationship between energy and momentum by algebraically manipulating their definitions. This produces $E^2 = (pc)^2 + (mc^2)^2$, where E is the relativistic total energy and p is the relativistic momentum. This relationship between relativistic energy and relativistic momentum is more complicated than the classical, but we can gain some interesting new insights by examining it. First, total energy is related to momentum and rest mass. At rest, momentum is zero, and the equation gives the total energy to be the rest energy mc^2 (so this equation is consistent with the discussion of rest energy above). However, as the mass is accelerated, its momentum p increases, thus increasing the total energy. At sufficiently high velocities, the rest energy term $(mc^2)^2$becomes negligible compared with the momentum term $(pc)^2$; thus, $E = pc$ at extremely relativistic velocities.

If we consider momentum p to be distinct from mass, we can determine the implications of the equation $E^2 = (pc)^2 + (mc^2)^2$, for a particle that has no mass. If we take m to be zero in this equation, then,

$$E = pc, or\ p = \frac{E}{c}.$$

Massless particles have this momentum. There are several massless particles found in nature, including photons (these are quanta of electromagnetic radiation). Another implication is that a massless particle must travel at speed c and only at speed c. While it is beyond the scope of this text to examine the relationship in the equation $E^2 = (pc)^2 + (mc^2)^2$, in detail, we can see that the relationship has important implications in special relativity.

Origins and Derivation of the Equation

The Energy–momentum relation was first established by Paul Dirac in 1928 under the form $E = \sqrt{(c^2p^2 + (m_0c^2)^2)} + V$, where V is the amount of potential energy.

The equation can be derived in a number of ways, two of the simplest include:

1. From the relativistic dynamics of a massive particle,
2. By evaluating the norm of the four-momentum of the system. This method applies to both massive and masseless particles, and can be extended to multi-particle systems with relatively little effort.

Heuristic Approach for Massive Particles

For a massive object moving at three-velocity u = (u_x, u_y, u_z) with magnitude |u| = u in the lab frame:

$$E = \gamma_{(\mathrm{u})} m_0 c^2$$

is the total energy of the moving object in the lab frame,

$$\mathrm{p} = \gamma_{(\mathrm{u})} m_0 \mathrm{u}$$

is the three dimensional relativistic momentum of the object in the lab frame with magnitude |p| = p. The relativistic energy E and momentum p include the Lorentz factor defined by:

$$\gamma_{(\mathrm{u})} = \frac{1}{\sqrt{1 - \frac{\mathrm{u} \cdot \mathrm{u}}{c^2}}} = \frac{1}{\sqrt{1 - \left(\frac{u}{c}\right)^2}}$$

Some authors use relativistic mass defined by:

$$m = \gamma_{(\mathrm{u})} m_0$$

although rest mass m_0 has a more fundamental significance, and will be used primarily over relativistic mass m.

Squaring the 3-momentum gives:

$$p^2 = \mathrm{p} \cdot \mathrm{p} = \frac{m_0^2 \mathrm{u} \cdot \mathrm{u}}{1 - \frac{\mathrm{u} \cdot \mathrm{u}}{c^2}} = \frac{m_0^2 u^2}{1 - \left(\frac{u}{c}\right)^2}$$

then solving for u^2 and substituting into the Lorentz factor obtains its alternative form in terms of 3-momentum and mass, rather than 3-velocity:

$$\gamma = \sqrt{1 + \left(\frac{p}{m_0 c}\right)^2}$$

Inserting this form of the Lorentz factor into the energy equation:

$$E = m_0 c^2 \sqrt{1 + \left(\frac{p}{m_0 c}\right)^2}$$

followed by more rearrangement yields (*1*). The elimination of the Lorentz factor also eliminates implicit velocity dependence of the particle in (*1*), as well as any inferences to the "relativistic mass" of a massive particle. This approach is not general as massless particles are not considered. Naively setting $m_0 = 0$ would mean that $E = 0$ and p = 0 and no energy–momentum relation could be derived, which is not correct.

Norm of the Four-momentum

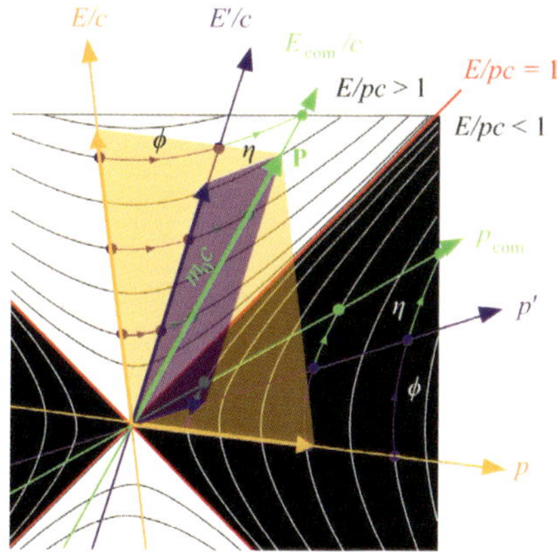

The energy and momentum of an object measured in two inertial frames in energy–momentum space – the yellow frame measures E and p while the blue frame measures E' and p′. The green arrow is the four-momentum P of an object with length proportional to its rest mass m_0. The green frame is the centre-of-momentum frame for the object with energy equal to the rest energy. The hyperbolae show the Lorentz transformation from one frame to another is a hyperbolic rotation, and ϕ and $\phi + \eta$ are the rapidities of the blue and green frames, respectively.

Special Relativity

In Minkowski space, energy (divided by c) and momentum are two components of a Minkowski four-vector, namely the four-momentum;

$$\mathrm{P} = \left(\frac{E}{c}, \mathrm{p}\right),$$

(these are the contravariant components).

The Minkowski inner product $\langle , \rangle$ of this vector with itself gives the square of the norm of this vector, it is proportional to the square of the rest mass m of the body:

$$\langle \mathrm{P}, \mathrm{P} \rangle = |\mathrm{P}|^2 = (m_0 c)^2,$$

a Lorentz invariant quantity, and therefore independent of the frame of reference. Using the Minkowski metric η with metric signature$(- + + +)$, the inner product is,

$$\langle \mathrm{P}, \mathrm{P} \rangle = |\mathrm{P}|^2 = -(m_0 c)^2,$$

and

$$\langle \mathrm{P}, \mathrm{P} \rangle = P^\alpha \eta_{\alpha\beta} P^\beta = \begin{pmatrix} \frac{E}{c} & p_x & p_y & p_z \end{pmatrix} \begin{pmatrix} -1 & 0 & 0 & 0 \\ 0 & 1 & 0 & 0 \\ 0 & 0 & 1 & 0 \\ 0 & 0 & 0 & 1 \end{pmatrix} \begin{pmatrix} \frac{E}{c} \\ p_x \\ p_y \\ p_z \end{pmatrix} = -\left(\frac{E}{c}\right)^2 + p^2,$$

so,

$$-(m_o c)^2 = -\left(\frac{E}{c}\right)^2 + p^2.$$

General Relativity

In general relativity, the 4-momentum is a four-vector defined in a local coordinate frame, although by definition the inner product is similar to that of special relativity,

$$\langle \mathrm{P,P} \rangle = |\mathrm{P}|^2 = (m_o c)^2,$$

in which the Minkowski metric η is replaced by the metric tensor field g:

$$\langle \mathrm{P,P} \rangle = |\mathrm{P}|^2 = P^\alpha g_{\alpha\beta} P^\beta,$$

solved from the Einstein field equations. Then:

$$P^\alpha g_{\alpha\beta} P^\beta = (m_o c)^2.$$

Performing the summations over indices followed by collecting "time-like", "spacetime-like", and "space-like" terms gives:

$$\underbrace{g_{oo}\left(P^o\right)^2}_{\text{time-like}} + 2\underbrace{g_{oi} P^o P^i}_{\text{spacetime-like}} + \underbrace{g_{ij} P^i P^j}_{\text{space-like}} = (m_o c)^2.$$

where the factor of 2 arises because the metric is a symmetric tensor, and the convention of Latin indices i, j taking space-like values 1, 2, 3 is used. As each component of the metric has space and time dependence in general; this is significantly more complicated than the formula quoted at the beginning.

Galilean Transformation

Galilean transformations, also called Newtonian transformations, set of equations in classical physics that relate the space and time coordinates of two systems moving at a constant velocityrelative to each other. Adequate to describe phenomena at speeds much smaller than the speed of light, Galilean transformations formally express the ideas that space and time are absolute; that length, time, and mass are independent of the relative motion of the observer; and that the speed of light depends upon the relative motion of the observer.

Suppose there are two reference frames (systems) designated by S and S' such that the co-ordinate axes are parallel. In S, we have the co-ordinates $\{x,y,z,t\}$ and in S' we have the co-ordinates $\{x',y',z',t'\}$. S' is moving with respect to S with velocity v (as measured in S) in the x direction. The clocks in both systems were synchronised at time $t = 0$ and they run at the same rate.

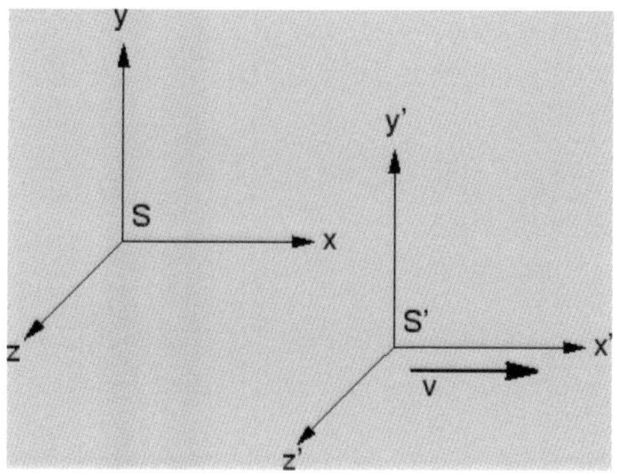

Reference frame S' moves with velocity v (in the x direction) relative to reference frame S

We have the intuitive relationships,

$$\begin{aligned} x' &= x - vt \\ y' &= y \\ z' &= z \\ t' &= t \end{aligned}$$

This set of equations is known as the Galilean Transformation. They enable us to relate a measurement in one inertial reference frame to another. For example, suppose we measure the velocity of a vehicle moving in the in x-direction in system S, and we want to know what would be the velocity of the vehicle in S'.

$$v'_x = \frac{dx'}{dt'} = \frac{d(x - vt)}{dt} = v_x - v$$

This is the result our intuition is familiar with.

We have stated the we would like the laws of physics to be the same in all inertial reference frames, as this is indeed our experience of nature. Physically, we should be able to perform the same experiments in different reference frames, and find always the same physical laws. Mathematically, these laws are expressed by equations. So, we should be able to "transform" our equations from one inertial reference frame to the other inertial reference frame, and always find the same answer.

Suppose we wanted to check that Newton's Second Law is the same in two different reference frames. We put one observer in the un-primed frame, and the other in the primed frame, moving with velocity v relative to the un-primed frame. Consider the vehicle of the previous case undergoing a constant acceleration in the x-direction,

$$\begin{aligned} f' = m'a' &= m'\frac{d^2x'}{dt'^2} \\ &= m'\frac{d}{dt'}\left(\frac{dx'}{dt'}\right) \\ &= m'\frac{d}{dt}\left(\frac{d(x - vt)}{dt}\right) \end{aligned}$$

$$= m\frac{d(v_x - v)}{dt}$$
$$= m\frac{dv_x}{dt}$$
$$= ma = f$$

Indeed, it does not matter which inertial frame we observe from, we recover the same Second Law of Motion each time. In the parlance of physics, we say the Second Law of Motion is invariant under the Galilean Transformation.

Exercise

You show that the Law of Momentum Conservation holds regardless of the inertial frame a given collision is viewed in. This is done by specialising to a collision where all velocities are in the x-direction. How would you do this for a more general collision ?

So far so good !

We have Classical Mechanics, a beautiful theory, as it has an elegant independence of how you observe it. There is a sense of poetry in how ugly terms, arising from observation in a different frame, eventually drop away, until we are left with physical laws which are invariant under the Galilean Transformation. But as time passes, it becomes clear we are in a fools paradise.

The first problem. Experiments on electric and magnetic fields, as well as induction of one type of field from changes in the other, lead to the collection of a set of equations, describing all these phenomena, known as Maxwell's Equations. You are already familiar with them. In vacuum they are,

MaxwellsEquations *in vacuo*

$$\nabla.\mathrm{B} = 0,$$
$$\nabla.\mathrm{E} = 0,$$
$$\nabla\times\mathrm{B} = \epsilon_0\,\mu_0\frac{\partial E}{\partial t},$$
$$\nabla\times\mathrm{E} = -\frac{\partial \mathrm{B}}{\partial t}.$$

Now, these equations are considered to be rock solid, arising from and verified by many experiments. Amazingly, they imply the existence of a previously not guessed at phenomenon. This is the electromagnetic wave. Every electrical engineer, following Marconi, must appreciate this !

To see this in detail, take the time derivative of the second last equation and the curl of the last.

$$\frac{\partial}{\partial t}\nabla\times B = \epsilon_0\,\mu_0\frac{\partial E}{\partial t},$$
$$\nabla\times(\nabla\times\mathrm{E}) = -\nabla\times\frac{\partial B}{\partial t}.$$

Now note that space and time derivatives commute,

$$\frac{\partial}{\partial t}\nabla\times\mathrm{B} = \nabla\times\frac{\partial B}{\partial t},$$

so,

$$\nabla \times (\nabla \times \mathrm{E}) = -\epsilon_0 \mu_0 \frac{\partial^2 E}{\partial t^2}.$$

Now, we use the identity,

$$\nabla \times (\nabla \times \mathrm{E}) = \nabla \nabla . \mathrm{E} - \nabla^2 \mathrm{E}.$$

The second term of the above equation drops out due to the vanishing of the divergence of the electric field (the second of Maxwell's Equations). So, we finally have the three dimensional wave equation,

$$\nabla^2 \mathrm{E} = \epsilon_0 \mu_0 \frac{\partial^2 E}{\partial t^2}.$$

To see this is a wave equation, note the analogy in one dimension,

$$\frac{\partial^2 y}{\partial x^2} = \frac{1}{c^2} \frac{\partial^2 y}{\partial t^2}.$$

which is solved by the wave function,

$$y(x,t) = \sin(x - ct),$$

which represents a wave traveling along the x axis with velocity c.

It is clear therefore that Maxwell's Equations are highly predictive.

1. A diversity is unified in a simplicity. The various phenomena of radiowaves, microwaves, infrared, visible and ultra-violet light, X-rays and gamma rays are all electromagnetic waves, differing only in their frequency.
2. They all travel at the same speed.
3. Even that speed is specified : $c = 1/\sqrt{\epsilon_0 \mu_0} = 2.997 \times 10^8$ m/s.
4. The speed appears independent of the source and the observer.

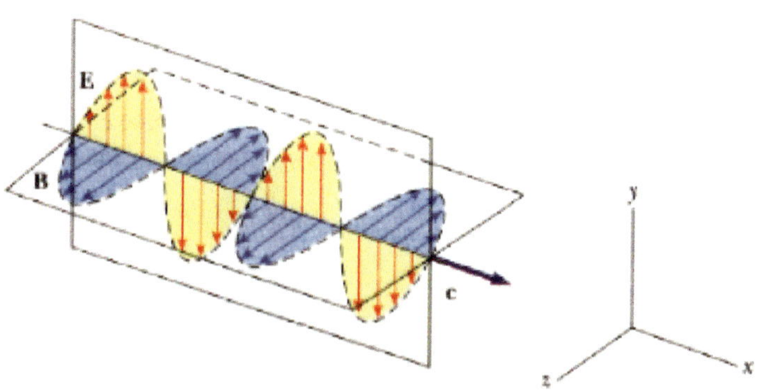

An electromagnetic wave traveling in vacuum

Lorentz Transformation

Lorentz transformations, set of equations in relativity physics that relate the space and time coordinates of two systems moving at a constant velocity relative to each other. Required to describe high-speed phenomena approaching the speed of light, Lorentz transformations formally express the relativity concepts that space and time are not absolute; that length, time, and mass depend on the relative motion of the observer; and that the speed of light in a vacuum is constant and independent of the motion of the observer or the source. The equations were developed by the Dutch physicist Hendrik Antoon Lorentz in 1904.

These co-ordinate transformations are known as the Lorentz Transformation. Note that the Lorentz Transformation reduces to the Galilean Transformation when $v \ll c$ so that $k = 1$ as required. Lorentz derived this transformation which made the Maxwell Equations of electromagnetism the same in all inertial reference frames in 1895. It was Einstein who is credited (in 1905) with discovering the full significance of these equations : The equations of the Lorentz Transformation define the relativity of space-time when measurements are made in different inertial reference frames.

The fact the the speed of light is "c" (constant) in all inertial reference frames implies a relativity in space-time. We expect length and time measurements to differ in different frames, in such a way that the speed of light is constant.

Because the relativistic effects become more obvious at velocities approaching the speed of light, a special symbol, "*beta*" defined as,

$$\beta = v / c$$

is used in relativistic parlance. Clearly, for velocities such that $t_2 = t_1$, relativistic effects are negligible, and for velocities such that $L = L_0\sqrt{1 - v^2 / c^2} = L_0 / \gamma$, relativistic effects must be considered. The scale of relativity between reference fames is also an important characteristic. This is given the the special symbol "*gamma*" defined as,

$$\gamma = \frac{1}{\sqrt{1 - v^2 / c^2}}$$

Clearly, non-relativistic velocities $t_2 = t_1$ imply $\gamma \approx 1$, whereas extreme relativistic velocities imply $\gamma \gg 1$, as in figure below.

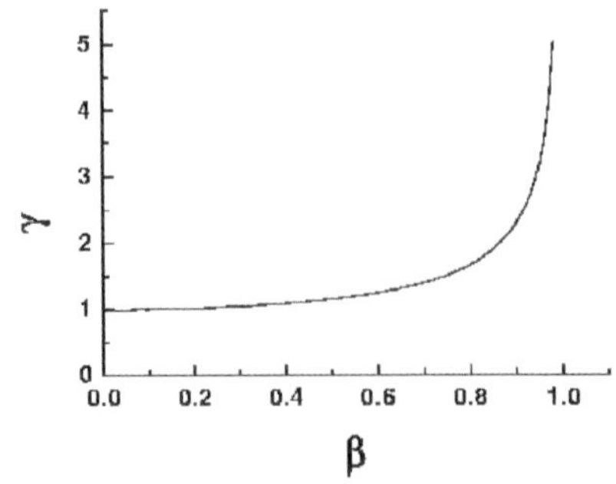

The gradual development of relativistic effects as $v \to c$

Derivation of the Group of Lorentz Transformations

An *event* is something that happens at a certain point in spacetime, or more generally, the point in spacetime itself. In any inertial frame an event is specified by a time coordinate *ct*and a set of Cartesian coordinates x, y, z to specify position in space in that frame. Subscripts label individual events.

From Einstein's second postulate of relativity follows,

$$c^2(t_2-t_1)^2-(x_2-x_1)^2-(y_2-y_1)^2-(z_2-z_1)^2=0 \quad \text{(lightlike separated events 1, 2)}$$

in all inertial frames for events connected by *light signals*. The quantity on the left is called the *spacetime interval* between events $a_1 = (t_1, x_1, y_1, z_1)$ and $a_2 = (t_2, x_2, y_2, z_2)$. The interval between *any two* events, not necessarily separated by light signals, is in fact invariant, i.e., independent of the state of relative motion of observers in different inertial frames, as is shown using homogeneity and isotropy of space. The transformation sought after thus must possess the property that,

$$\begin{aligned} &c^2(t_2-t_1)^2-(x_2-x_1)^2-(y_2-y_1)^2-(z_2-z_1)^2 \\ &=c^2(t_{2'}-t_{1'})^2-(x_{2'}-x_{1'})^2-(y_{2'}-y_{1'})^2-(z_{2'}-z_{1'})^2 \quad \text{(all events 1, 2).} \end{aligned}$$

where (ct, x, y, z) are the spacetime coordinates used to define events in one frame, and (ct', x', y', z') are the coordinates in another frame. First one observes that equation above is satisfied if an arbitrary 4-tuple b of numbers are added to events a_1 and a_2. Such transformations are called *spacetime translations* and are not dealt with further here. Then one observes that a *linear* solution preserving the origin of the simpler problem,

$$\begin{aligned} &c^2t^2-x^2-y^2-z^2=c^2t'^2-x'^2-y'^2-z'^2 \\ \text{or}\quad &c^2t_1t_2-x_1x_2-y_1y_2-z_1z_2=c^2t_1't_2'-x_1'x_2'-y_1'y_2'-z_1'z_2' \end{aligned}$$

solves the general problem too. Finding the solution to the simpler problem is just a matter of look-up in the theory of classical groups that preserve bilinear forms of various signature.Equation above can be written more compactly as,

$$(a,a)=(a',a') \quad \text{or} \quad a\cdot a=a'\cdot a',$$

where $(\cdot, \cdot)$ refers to the bilinear form of signature (1, 3) on R^4 exposed by the right hand side formula in third set of equations. The alternative notation defined on the right is referred to as the *relativistic dot product*. Spacetime mathematically viewed as R^4 endowed with this bilinear form is known as Minkowski space M. The Lorentz transformation is thus an element of the group Lorentz group O(1, 3), the Lorentz group or, for those that prefer the other metric signature, O(3, 1) (also called the Lorentz group). One has,

$$(a,a)=(\Lambda a,\Lambda a)=(a',a'), \quad \Lambda\in O(1,3), \quad a,a'\in M,$$

which is precisely preservation of the bilinear form in third set of equations which implies (by linearity of Λ and bilinearity of the form) that second set of equations is satisfied. The elements of the Lorentz group are rotations and *boosts* and mixes thereof. If the spacetime translations are included, then one obtains the *inhomogeneous Lorentz group* or the Poincaré group.

Generalities

The relations between the primed and unprimed spacetime coordinates are the Lorentz transformations, each coordinate in one frame is a linear function of all the coordinates in the other frame, and the inverse functions are the inverse transformation. Depending on how the frames move relative to each other, and how they are oriented in space relative to each other, other parameters that describe direction, speed, and orientation enter the transformation equations.

Transformations describing relative motion with constant (uniform) velocity and without rotation of the space coordinate axes are called *boosts*, and the relative velocity between the frames is the parameter of the transformation. The other basic type of Lorentz transformations is rotations in the spatial coordinates only, these are also inertial frames since there is no relative motion, the frames are simply tilted (and not continuously rotating), and in this case quantities defining the rotation are the parameters of the transformation (e.g., axis–angle representation, or Euler angles, etc.). A combination of a rotation and boost is a *homogeneous transformation*, which transforms the origin back to the origin.

The full Lorentz group O(3, 1) also contains special transformations that are neither rotations nor boosts, but rather reflections in a plane through the origin. Two of these can be singled out; spatial inversion in which the spatial coordinates of all events are reversed in sign and temporal inversion in which the time coordinate for each event gets its sign reversed.

Boosts should not be conflated with mere displacements in spacetime; in this case, the coordinate systems are simply shifted and there is no relative motion. However, these also count as symmetries forced by special relativity since they leave the spacetime interval invariant. A combination of a rotation with a boost, followed by a shift in spacetime, is an *inhomogeneous Lorentz transformation*, an element of the Poincaré group, which is also called the inhomogeneous Lorentz group.

Physical Formulation of Lorentz Boosts

Coordinate Transformation

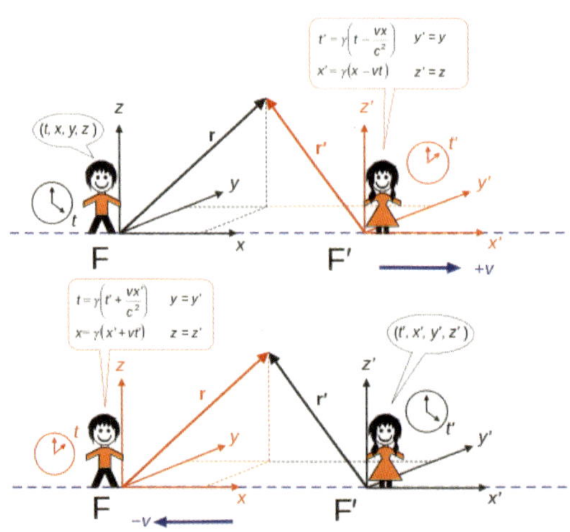

The spacetime coordinates of an event, as measured by each observer in their inertial reference frame (in standard configuration) are shown in the speech bubbles.
Top: frame F' moves at velocity v along the x-axis of frame F.
Bottom: frame F moves at velocity $-v$ along the x'-axis of frame F'

A "stationary" observer in frame F defines events with coordinates t, x, y, z. Another frame F' moves with velocity v relative to F, and an observer in this "moving" frame F' defines events using the coordinates t', x', y', z'.

The coordinate axes in each frame are parallel (the x and x' axes are parallel, the y and y' axes are parallel, and the z and z'axes are parallel), remain mutually perpendicular, and relative motion is along the coincident xx' axes. At $t = t' = 0$, the origins of both coordinate systems are the same, $(x, y, z) = (x', y', z') = (0, 0, 0)$. In other words, the times and positions are coincident at this event. If all these hold, then the coordinate systems are said to be in standard configuration, or synchronized.

If an observer in F records an event t, x, y, z, then an observer in F' records the *same* event with coordinates.

Lorentz boost (*x direction*):

$$t' = \gamma\left(t - \frac{vx}{c^2}\right)$$
$$x' = \gamma\left(x - vt\right)$$
$$y' = y$$
$$z' = z$$

where v is the relative velocity between frames in the x-direction, c is the speed of light, and

$$\gamma = \frac{1}{\sqrt{1 - \frac{v^2}{c^2}}}$$

(lowercase gamma) is the Lorentz factor.

Here, v is the *parameter* of the transformation, for a given boost it is a constant number, but can take a continuous range of values. In the setup used here, positive relative velocity $v > 0$ is motion along the positive directions of the xx' axes, zero relative velocity $v = 0$ is no relative motion, while negative relative velocity $v < 0$ is relative motion along the negative directions of the xx' axes. The magnitude of relative velocity v cannot equal or exceed c, so only subluminal speeds $-c < v < c$ are allowed. The corresponding range of γ is $1 \le \gamma < \infty$.

The transformations are not defined if v is outside these limits. At the speed of light ($v = c$) γ is infinite, and faster than light ($v > c$) γ is a complex number, each of which make the transformations unphysical. The space and time coordinates are measurable quantities and numerically must be real numbers.

As an active transformation, an observer in F′ notices the coordinates of the event to be "boosted" in the negative directions of the xx' axes, because of the $-v$ in the transformations. This has the equivalent effect of the *coordinate system* F′ boosted in the positive directions of the xx' axes, while the event does not change and is simply represented in another coordinate system, a passive transformation.

The inverse relations (t, x, y, z in terms of t', x', y', z') can be found by algebraically solving the original set of equations. A more efficient way is to use physical principles. Here F' is the "stationary" frame while F is the "moving" frame. According to the principle of relativity, there is no privileged frame of reference, so the transformations from F' to F must take exactly the same form as the transformations from F to F'. The only difference is F moves with velocity $-v$ relative to F' (i.e., the relative velocity has the same magnitude but is oppositely directed). Thus if an observer in F' notes an event t', x', y', z', then an observer in F notes the *same* event with coordinates.

Inverse Lorentz boost (x direction):

$$t = \gamma\left(t' + \frac{vx'}{c^2}\right)$$
$$x = \gamma(x' + vt')$$
$$y = y'$$
$$z = z',$$

and the value of γ remains unchanged. This "trick" of simply reversing the direction of relative velocity while preserving its magnitude, and exchanging primed and unprimed variables, always applies to finding the inverse transformation of every boost in any direction.

Sometimes it is more convenient to use $\beta = v/c$ (lowercase beta) instead of v, so that,

$$ct' = \gamma(ct - \beta x),$$
$$x' = \gamma(x - \beta ct),$$

which shows much more clearly the symmetry in the transformation. From the allowed ranges of v and the definition of β, it follows $-1 < \beta < 1$. The use of β and γ is standard throughout the literature.

The Lorentz transformations can also be derived in a way that resembles circular rotations in 3d space using the hyperbolic functions. For the boost in the x direction, the results are,

Lorentz boost (x *direction with rapidity* ζ):

$$ct' = ct\cosh\zeta - x\sinh\zeta$$
$$x' = x\cosh\zeta - ct\sinh\zeta$$
$$y' = y$$
$$z' = z$$

where ζ (lowercase zeta) is a parameter called *rapidity* (many other symbols are used, including θ, ϕ, φ, η, ψ, ξ). Given the strong resemblance to rotations of spatial coordinates in 3d space in the Cartesian xy, yz, and zx planes, a Lorentz boost can be thought of as a hyperbolic rotation of spacetime coordinates in the xt, yt, and zt Cartesian-time planes of 4d Minkowski space. The parameter ζ is the hyperbolic angle of rotation, analogous to the ordinary angle for circular rotations. This transformation can be illustrated with a Minkowski diagram.

The hyperbolic functions arise from the *difference* between the squares of the time and spatial coordinates in the spacetime interval, rather than a sum. The geometric significance of the hyperbolic functions can be visualized by taking $x = 0$ or $ct = 0$ in the transformations. Squaring and subtracting the results, one can derive hyperbolic curves of constant coordinate values but varying ζ, which parametrizes the curves according to the identity,

$$cosh^2 \zeta - sinh^2 \zeta = 1.$$

Conversely the ct and x axes can be constructed for varying coordinates but constant ζ. The definition,

$$tanh\,\zeta = \frac{sinh\,\zeta}{cosh\,\zeta},$$

provides the link between a constant value of rapidity, and the slope of the ct axis in spacetime. A consequence these two hyperbolic formulae is an identity that matches the Lorentz factor,

$$cosh\,\zeta = \frac{1}{\sqrt{1 - tanh^2 \zeta}}.$$

Comparing the Lorentz transformations in terms of the relative velocity and rapidity, or using the above formulae, the connections between β, γ, and ζ are,

$$\beta = tanh\,\zeta,$$
$$\gamma = cosh\,\zeta,$$
$$\beta\gamma = sinh\,\zeta.$$

Taking the inverse hyperbolic tangent gives the rapidity,

$$\zeta = tanh^{-1} \beta.$$

Since $-1 < \beta < 1$, it follows $-\infty < \zeta < \infty$. From the relation between ζ and β, positive rapidity $\zeta > 0$ is motion along the positive directions of the xx' axes, zero rapidity $\zeta = 0$ is no relative motion, while negative rapidity $\zeta < 0$ is relative motion along the negative directions of the xx' axes.

The inverse transformations are obtained by exchanging primed and unprimed quantities to switch the coordinate frames, and negating rapidity $\zeta \to -\zeta$ since this is equivalent to negating the relative velocity. Therefore,

Inverse Lorentz boost (*x direction with rapidity ζ*):

$$ct = ct' cosh\,\zeta + x' sinh\,\zeta$$
$$x = x' cosh\,\zeta + ct' sinh\,\zeta$$

$$y = y'$$
$$z = z'$$

The inverse transformations can be similarly visualized by considering the cases when $x' = 0$ and $ct' = 0$.

So far the Lorentz transformations have been applied to *one event*. If there are two events, there is a spatial separation and time interval between them. It follows from the linearity of the Lorentz transformations that two values of space and time coordinates can be chosen, the Lorentz transformations can be applied to each, then subtracted to get the Lorentz transformations of the differences;

$$\Delta t' = \gamma\left(\Delta t - \frac{v\Delta x}{c^2}\right),$$

$$\Delta x' = \gamma\left(\Delta x - v\Delta t\right),$$

with inverse relations,

$$\Delta t = \gamma\left(\Delta t' + \frac{v\Delta x'}{c^2}\right),$$

$$\Delta x = \gamma\left(\Delta x' + v\Delta t'\right).$$

where Δ (uppercase delta) indicates a difference of quantities; e.g., $\Delta x = x_2 - x_1$ for two values of x coordinates, and so on.

These transformations on *differences* rather than spatial points or instants of time are useful for a number of reasons:

- In calculations and experiments, it is lengths between two points or time intervals that are measured or of interest (e.g., the length of a moving vehicle, or time duration it takes to travel from one place to another),
- The transformations of velocity can be readily derived by making the difference infinitesimally small and dividing the equations, and the process repeated for the transformation of acceleration,
- If the coordinate systems are never coincident (i.e., not in standard configuration), and if both observers can agree on an event t_0, x_0, y_0, z_0 in F and t_0', x_0', y_0', z_0' in F', then they can use that event as the origin, and the spacetime coordinate differences are the differences between their coordinates and this origin, e.g., $\Delta x = x - x_0$, $\Delta x' = x' - x_0'$, etc.

Physical Implications

A critical requirement of the Lorentz transformations is the invariance of the speed of light, a fact used in their derivation, and contained in the transformations themselves. If in F the equation for a pulse of light along the x direction is $x = ct$, then in F' the Lorentz transformations give $x' = ct'$, and vice versa, for any $-c < v < c$.

For relative speeds much less than the speed of light, the Lorentz transformations reduce to the Galilean transformation.

$$t' \approx t$$

$$x' \approx x - vt$$

in accordance with the correspondence principle. It is sometimes said that nonrelativistic physics is a physics of "instantaneous action at a distance".

Three unintuitive, but correct, predictions of the transformations are mentioned below.

Time Dilation

Suppose there is a clock at rest in F. If a time interval (say a "tick") is measured at the same point so that $\Delta x = 0$, then the transformations give this tick in F' by $\Delta t' = \gamma\Delta t$. Conversely, suppose there is a clock at rest in F'. If a tick is measured at the same point so that $\Delta x' = 0$, then the transformations give this tick in F by $\Delta t = \gamma\Delta t'$. Either way, the boosted observer measures longer time intervals than the observer in the other frame.

Relativity of Simultaneity

Suppose two events occur simultaneously ($\Delta t = 0$) along the x axis, but separated by a nonzero displacement Δx. Then in F', we find that $\Delta t' = \gamma\frac{-v\Delta x}{c^2}$, so the events are no longer simultaneous according to a moving observer.

Length Contraction

Suppose there is a rod at rest in F aligned along the x axis, with length Δx. In F', the rod moves with velocity $-v$, so its length must be measured by taking two simultaneous ($\Delta t' = 0$) measurements at opposite ends. Under these conditions, the inverse Lorentz transform shows that $\Delta x = \gamma\Delta x'$. In F the two measurements are no longer simultaneous, but this does not matter because the rod is at rest in F. We conclude that the boosted observer measures a shorter length, by a factor of γ, than the observer in the rest frame of the rod. Length contraction affects any geometric quantity related to lengths, so from the perspective of a moving observer, areas and volumes will also appear to shrink along the direction of motion.

Vector Transformations

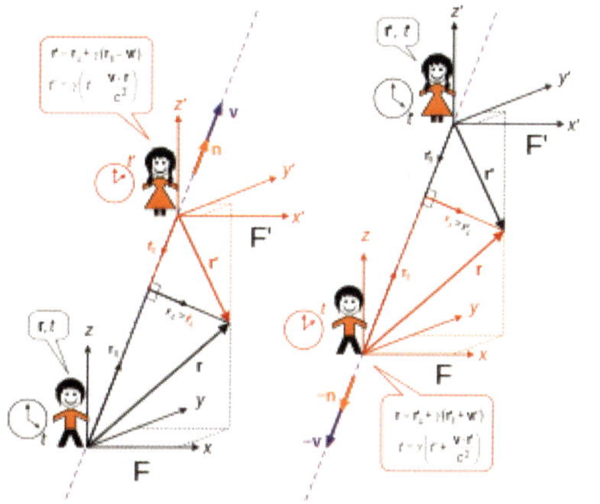

An observer in frame F observes F' to move with velocity v, while F' observes F to move with velocity −v. The coordinate axes of each frame are still parallel and orthogonal. The position vector as measured in each frame is split into components parallel and perpendicular to the relative velocity vector v. Left: Standard configuration. Right: Inverse configuration

The use of vectors allows positions and velocities to be expressed in arbitrary directions compactly. A single boost in any direction depends on the full relative velocity vector v with a magnitude $|\mathrm{v}| = \upsilon$ that cannot equal or exceed c, so that $0 \le \upsilon < c$.

Only time and the coordinates parallel to the direction of relative motion change, while those coordinates perpendicular do not. With this in mind, split the spatial position vector r as measured in F, and r′ as measured in F', each into components perpendicular (⊥) and parallel (||) to v,

$$\mathrm{r} = \mathrm{r}_\perp + \mathrm{r}_\parallel, \quad \mathrm{r}' = \mathrm{r}_{\perp'} + \mathrm{r}_{\parallel'},$$

then the transformations are,

$$t' = \gamma\left(t - \frac{\mathrm{r}_\parallel \cdot \mathrm{v}}{c^2}\right)$$

$$\mathrm{r}_{\parallel'} = \gamma(\mathrm{r}_\parallel - \mathrm{v}t)$$

$$\mathrm{r}_{\perp'} = \mathrm{r}_\perp$$

where · is the dot product. The Lorentz factor γ retains its definition for a boost in any direction, since it depends only on the magnitude of the relative velocity. The definition $\boldsymbol{\beta} = \mathrm{v}/c$ with magnitude $0 \le \beta < 1$ is also used by some authors.

Introducing a unit vector $\mathrm{n} = \mathrm{v}/\upsilon = \boldsymbol{\beta}/\beta$ in the direction of relative motion, the relative velocity is $\mathrm{v} = \upsilon\mathrm{n}$ with magnitude υ and direction n, and vector projection and rejection give respectively,

$$\mathrm{r}_\parallel = (\mathrm{r}\cdot\mathrm{n})\mathrm{n}, \quad \mathrm{r}_\perp = \mathrm{r} - (\mathrm{r}\cdot\mathrm{n})\mathrm{n}$$

Accumulating the results gives the full transformations,

Lorentz boost (*in direction* n *with magnitude* υ):

$$t' = \gamma\left(t - \frac{\upsilon\mathrm{n}\cdot\mathrm{r}}{c^2}\right),$$

$$\mathrm{r}' = \mathrm{r} + (\gamma - 1)(\mathrm{r}\cdot\mathrm{n})\mathrm{n} - \gamma t\upsilon\mathrm{n}.$$

The projection and rejection also applies to r′. For the inverse transformations, exchange r and r′ to switch observed coordinates, and negate the relative velocity $\mathrm{v} \to -\mathrm{v}$ (or simply the unit vector $\mathrm{n} \to -\mathrm{n}$ since the magnitude υ is always positive) to obtain,

Inverse Lorentz boost (*in direction* n *with magnitude* υ):

$$t = \gamma\left(t' + \frac{\mathrm{r}'\cdot\upsilon\mathrm{n}}{c^2}\right),$$

$$\mathrm{r} = \mathrm{r}' + (\gamma - 1)(\mathrm{r}'\cdot\mathrm{n})\mathrm{n} + \gamma t'\upsilon\mathrm{n},$$

The unit vector has the advantage of simplifying equations for a single boost, allows either v or $\boldsymbol{\beta}$ to be reinstated when convenient, and the rapidity parametrization is immediately obtained by replacing β and $\beta\gamma$. It is not convenient for multiple boosts.

The vectorial relation between relative velocity and rapidity is,

$$\hat{a} = \beta \mathrm{n} = \mathrm{n}\, tanh\, \zeta,$$

and the "rapidity vector" can be defined as,

$$æ = \zeta \mathrm{n} = \mathrm{n}\, tanh^{-1} \beta,$$

each of which serves as a useful abbreviation in some contexts. The magnitude of ζ is the absolute value of the rapidity scalar confined to $0 \le \zeta < \infty$, which agrees with the range $0 \le \beta < 1$.

Transformation of Velocities

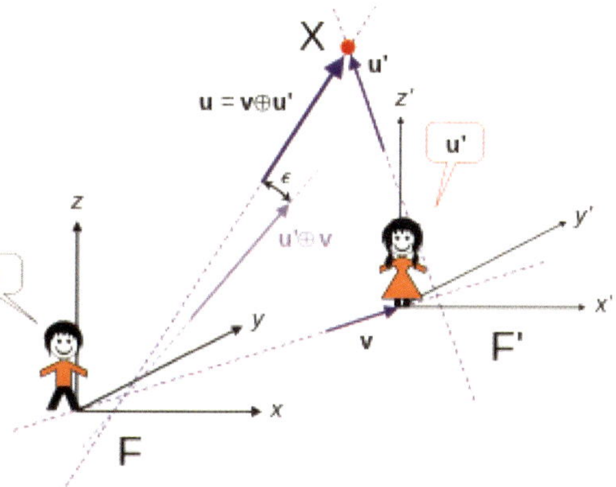

The transformation of velocities provides the definition relativistic velocity addition ⊕, the ordering of vectors is chosen to reflect the ordering of the addition of velocities; first v (the velocity of F′ relative to F) then **u′** (the velocity of X relative to F′) to obtain u = v ⊕ u′ (the velocity of X relative to F)

Defining the coordinate velocities and Lorentz factor by,

$$\mathrm{u} = \frac{dr}{dt}, \quad \mathrm{u}' = \frac{dr'}{dt'}, \quad \gamma_{\mathrm{v}} = \frac{1}{\sqrt{1 - \frac{\mathrm{v} \cdot \mathrm{v}}{c^2}}}$$

taking the differentials in the coordinates and time of the vector transformations, then dividing equations, leads to,

$$\mathrm{u}' = \frac{1}{1 - \frac{\mathrm{v} \cdot \mathrm{u}}{c^2}} \left[\frac{\mathrm{u}}{\gamma_{\mathrm{v}}} - \mathrm{v} + \frac{1}{c^2} \frac{\gamma_{\mathrm{v}}}{\gamma_{\mathrm{v}} + 1} (\mathrm{u} \cdot \mathrm{v}) \mathrm{v} \right]$$

The velocities u and u′ are the velocity of some massive object. They can also be for a third inertial frame (say F''), in which case they must be *constant*. Denote either entity by X. Then X moves with velocity u relative to F, or equivalently with velocity u′relative to F′, in turn F′ moves with velocity v relative to F. The inverse transformations can be obtained in a similar way, or as with position coordinates exchange u and u′, and change B to −v.

The transformation of velocity is useful in stellar aberration, the Fizeau experiment, and the relativistic Doppler effect.

The Lorentz transformations of acceleration can be similarly obtained by taking differentials in the velocity vectors, and dividing these by the time differential.

Transformation of other Quantities

In general, given four quantities A and Z = (Z_x, Z_y, Z_z) and their Lorentz-boosted counterparts A' and Z′ = (Z'_x, Z'_y, Z'_z), a relation of the form,

$$A^2 - \mathrm{Z}\cdot\mathrm{Z} = A'^2 - \mathrm{Z}'\cdot\mathrm{Z}'$$

implies the quantities transform under Lorentz transformations similar to the transformation of spacetime coordinates;

$$A' = \gamma\left(A - \frac{\upsilon\mathrm{n}\cdot\mathrm{Z}}{c}\right),$$

$$\mathrm{Z}' = \mathrm{Z} + (\gamma - 1)(\mathrm{Z}\cdot\mathrm{n})\mathrm{n} - \frac{\gamma A \upsilon \mathrm{n}}{c}.$$

The decomposition of Z (and Z′) into components perpendicular and parallel to v is exactly the same as for the position vector, as is the process of obtaining the inverse transformations (exchange (A, Z) and (A', Z′) to switch observed quantities, and reverse the direction of relative motion by the substitution n $\mapsto$ −n).

The quantities (A, Z) collectively make up a *four vector*, where A is the "timelike component", and Z the "spacelike component". Examples of A and Z are the following:

Four vector	A	Z
Position four vector	Time (multiplied by c), ct	Position vector, r
Four momentum	Energy (divided by c), E/c	Momentum, p
Four-wave vector	angular frequency (divided by c), ω/c	wave vector, k
Four spin	(No name), s_t	Spin, s
Four current	Charge density (multiplied by c), ρc	Current density, j
Electromagnetic four potential	Electric potential (divided by c), φ/c	Magnetic potential, A

For a given object (e.g., particle, fluid, field, material), if A or Z correspond to properties specific to the object like its charge density, mass density, spin, etc., its properties can be fixed in the rest frame of that object. Then the Lorentz transformations give the corresponding properties in a frame moving relative to the object with constant velocity. This breaks some notions taken for granted in non-relativistic physics. For example, the energy E of an object is a scalar in non-relativistic mechanics, but not in relativistic mechanics because energy changes under Lorentz transformations; its value is different for various inertial frames. In the rest frame of an object, it has a rest energy and zero momentum. In a boosted frame its energy is different and it appears to have a momentum. Similarly, in non-relativistic quantum mechanics the spin of a particle is a constant vector, but in relativistic quantum mechanics spin s depends on relative motion. In the rest frame of the particle, the spin pseudovector can be fixed to be its ordinary non-relativistic spin with a zero timelike quantity s_t, however a boosted observer will perceive a nonzero timelike component and an altered spin.

Not all quantities are invariant in the form as shown above, for example orbital angular momentum L does not have a timelike quantity, and neither does the electric field E nor the magnetic field B. The definition of angular momentum is L = r × p, and in a boosted frame the altered angular momentum is L′ = r′ × p′. Applying this definition using the transformations of coordinates and momentum leads to the transformation of angular momentum. It turns out L transforms with another vector quantity N = (E/c^2)r − tp related to boosts. For the case of the E and B fields, the transformations cannot be obtained as directly using vector algebra. The Lorentz force is the definition of these fields, and in F it is F = q(E + v × B) while in F' it is F′ = q(E′ + v′ × B′).

Tensor Formulation

Contravariant Vectors

Writing the general matrix transformation of coordinates as the matrix equation,

$$\begin{bmatrix} x'^0 \\ x'^1 \\ x'^2 \\ x'^3 \end{bmatrix} = \begin{bmatrix} \Lambda^0{}_0 & \Lambda^0{}_1 & \Lambda^0{}_2 & \Lambda^0{}_3 \\ \Lambda^1{}_0 & \Lambda^1{}_1 & \Lambda^1{}_2 & \Lambda^1{}_3 \\ \Lambda^2{}_0 & \Lambda^2{}_1 & \Lambda^2{}_2 & \Lambda^2{}_3 \\ \Lambda^3{}_0 & \Lambda^3{}_1 & \Lambda^3{}_2 & \Lambda^3{}_3 \end{bmatrix} \begin{bmatrix} x^0 \\ x^1 \\ x^2 \\ x^3 \end{bmatrix}$$

allows the transformation of other physical quantities that cannot be expressed as four-vectors; e.g., tensors or spinors of any order in 4d spacetime, to be defined. In the corresponding tensor index notation, the above matrix expression is,

$$x'^\nu = \Lambda^\nu{}_\mu x^\mu,$$

where lower and upper indices label covariant and contravariant components respectively, and the summation convention is applied. It is a standard convention to use Greekindices that take the value 0 for time components, and 1, 2, 3 for space components, while Latin indices simply take the values 1, 2, 3, for spatial components. Note that the first index (reading left to right) corresponds in the matrix notation to a *row index*. The second index corresponds to the column index.

The transformation matrix is universal for all four-vectors, not just 4-dimensional spacetime coordinates. If A is any four-vector, then in tensor index notation,

$$A'^\nu = \Lambda^\nu{}_\mu A^\mu.$$

Alternatively, one writes,

$$A^{\nu'} = \Lambda^{\nu'}{}_\mu A^\mu.$$

in which the primed indices denote the indices of A in the primed frame. This notation cuts risk of exhausting the Greek alphabet roughly in half.

For a general n-component object one may write,

$$X'^\alpha = \Pi(\Lambda)^\alpha{}_\beta X^\beta,$$

where Π is the appropriate representation of the Lorentz group, an $n \times n$ matrix for every Λ. In this case, the indices should *not* be thought of as spacetime indices (sometimes called Lorentz indices), and they run from 1 to n. E.g., if X is a bispinor, then the indices are called *Dirac indices*.

Covariant Vectors

There are also vector quantities with covariant indices. They are generally obtained from their corresponding objects with contravariant indices by the operation of *lowering an index*; e.g.,

$$x_\nu = \eta_{\mu\nu} x^\mu,$$

where η is the metric tensor. The inverse of this transformation is given by,

$$x^\mu = \eta^{\nu\mu} x_\nu,$$

where, when viewed as matrices, $\eta^{\mu\nu}$ is the inverse of $\eta_{\mu\nu}$. As it happens, $\eta^{\mu\nu} = \eta_{\mu\nu}$. This is referred to as *raising an index*. To transform a covariant vector A_μ, first raise its index, then transform it according to the same rule as for contravariant 4-vectors, then finally lower the index;

$$A'_\nu = \eta_{\rho\nu} \Lambda^\rho{}_\sigma \eta^{\mu\sigma} A_\mu.$$

But

$$\eta_{\rho\nu} \Lambda^\rho{}_\sigma \eta^{\mu\sigma} = \left(\Lambda^{-1}\right)^\mu{}_\nu,$$

i.e., it is the (μ, ν)-component of the *inverse* Lorentz transformation. One defines (as a matter of notation),

$$\Lambda_\nu{}^\mu \equiv \left(\Lambda^{-1}\right)^\mu{}_\nu,$$

and may in this notation write,

$$A'_\nu = \Lambda_\nu{}^\mu A_\mu.$$

Now for a subtlety. The implied summation on the right hand side of,

$$A'_\nu = \Lambda_\nu{}^\mu A_\mu = \left(\Lambda^{-1}\right)^\mu{}_\nu A_\mu$$

is running over *a row index* of the matrix representing Λ^{-1}. Thus, in terms of matrices, this transformation should be thought of as the *inverse transpose* of Λ acting on the column vector A_μ. That is, in pure matrix notation,

$$A' = \left(\Lambda^{-1}\right)^{\mathrm{T}} A.$$

This means exactly that covariant vectors (thought of as column matrices) transform according to the dual representation of the standard representation of the Lorentz group. This notion generalizes to general representations, simply replace Λ with Π(Λ).

Tensors

If A and B are linear operators on vector spaces U and V, then a linear operator $A \otimes B$ may be defined on the tensor product of U and V, denoted $U \otimes V$ according to,

$$(A \otimes B)(u \otimes v) = Au \otimes Bv, \qquad u \in U, v \in V, u \otimes v \in U \otimes V.$$

From this it is immediately clear that if u and v are a four-vectors in V, then $u \otimes v \in T_2 V \equiv V \otimes V$ transforms as,

$$u \otimes v \rightarrow \Lambda u \otimes \Lambda v = \Lambda^{\mu}{}_{\nu} u^{\nu} \otimes \Lambda^{\rho}{}_{\sigma} v^{\sigma} = \Lambda^{\mu}{}_{\nu} \Lambda^{\rho}{}_{\sigma} u^{\nu} \otimes v^{\sigma} \equiv \Lambda^{\mu}{}_{\nu} \Lambda^{\rho}{}_{\sigma} w^{\nu\sigma}.$$

The second step uses the bilinearity of the tensor product and the last step defines a 2-tensor on component form, or rather, it just renames the tensor $u \otimes v$.

These observations generalize in an obvious way to more factors, and using the fact that a general tensor on a vector space V can be written as a sum of a coefficient (component!) times tensor products of basis vectors and basis covectors, one arrives at the transformation law for any tensor quantity T. It is given by,

$$T^{\alpha'\beta'\cdots\zeta'}_{\theta'\iota'\cdots\kappa'} = \Lambda^{\alpha'}{}_{\mu} \Lambda^{\beta'}{}_{\nu} \cdots \Lambda^{\zeta'}{}_{\rho} \Lambda_{\theta'}{}^{\sigma} \Lambda_{\iota'}{}^{\upsilon} \cdots \Lambda_{\kappa'}{}^{\zeta} T^{\mu\nu\cdots\rho}_{\sigma\upsilon\cdots\zeta},$$

where $\Lambda_{\chi'}{}^{\psi}$ is defined above. This form can generally be reduced to the form for general n-component objects given above with a single matrix ($\Pi(\Lambda)$) operating on column vectors. This latter form is sometimes preferred; e.g., for the electromagnetic field tensor.

Transformation of the Electromagnetic Field

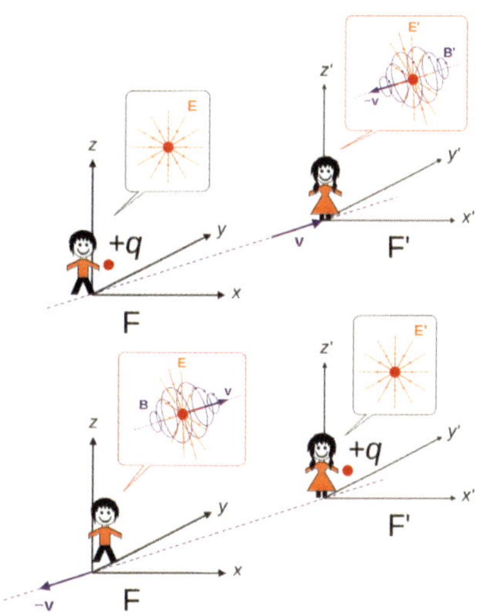

Lorentz boost of an electric charge, the charge is at rest in one frame or the other

Lorentz transformations can also be used to illustrate that the magnetic field B and electric field E are simply different aspects of the same force — the electromagnetic force, as a consequence of relative motion between electric charges and observers.The fact that the electromagnetic field shows relativistic effects becomes clear by carrying out a simple thought experiment.

- An observer measures a charge at rest in frame F. The observer will detect a static electric field. As the charge is stationary in this frame, there is no electric current, so the observer does not observe any magnetic field.
- The other observer in frame F′ moves at velocity v relative to F and the charge. *This* observer sees a different electric field because the charge moves at velocity −v in their rest frame. The motion of the charge corresponds to an electric current, and thus the observer in frame F′ also sees a magnetic field.

The electric and magnetic fields transform differently from space and time, but exactly the same way as relativistic angular momentum and the boost vector.

The electromagnetic field strength tensor is given by,

$$F^{\mu\nu} = \begin{bmatrix} 0 & -\frac{1}{c}E_x & -\frac{1}{c}E_y & -\frac{1}{c}E_z \\ \frac{1}{c}E_x & 0 & -B_z & B_y \\ \frac{1}{c}E_y & B_z & 0 & -B_x \\ \frac{1}{c}E_z & -B_y & B_x & 0 \end{bmatrix} \text{(SI units, signature } (+,-,-,-)).$$

in SI units. In relativity, the Gaussian system of units is often preferred over SI units, even in texts whose main choice of units is SI units, because in it the electric field E and the magnetic induction B have the same units making the appearance of the electromagnetic field tensor more natural. Consider a Lorentz boost in the x-direction. It is given by,

$$\Lambda^{\mu}{}_{\nu} = \begin{bmatrix} \gamma & -\gamma\beta & 0 & 0 \\ -\gamma\beta & \gamma & 0 & 0 \\ 0 & 0 & 1 & 0 \\ 0 & 0 & 0 & 1 \end{bmatrix}, \qquad F^{\mu\nu} = \begin{bmatrix} 0 & E_x & E_y & E_z \\ -E_x & 0 & B_z & -B_y \\ -E_y & -B_z & 0 & B_x \\ -E_z & B_y & -B_x & 0 \end{bmatrix} \text{(Gaussian units, signature } (-,+,+,+)),$$

where the field tensor is displayed side by side for easiest possible reference in the manipulations below.

The general transformation law $T^{\alpha'\beta'\cdots\zeta'}_{\theta'\iota'\cdots\kappa'} = \Lambda^{\alpha'}{}_{\mu}\Lambda^{\beta'}{}_{\nu}\cdots\Lambda^{\zeta'}{}_{\rho}\Lambda_{\theta'}{}^{\sigma}\Lambda_{\iota'}{}^{\upsilon}\cdots\Lambda_{\kappa'}{}^{\zeta}T^{\mu\nu\cdots\rho}_{\sigma\upsilon\cdots\zeta}$, becomes,

$$F^{\mu'\nu'} = \Lambda^{\mu'}{}_{\mu}\Lambda^{\nu'}{}_{\nu}F^{\mu\nu}.$$

For the magnetic field one obtains,

$$\begin{aligned} B_{x'} &= F^{2'3'} = \Lambda^{2}{}_{\mu}\Lambda^{3}{}_{\nu}F^{\mu\nu} = \Lambda^{2}{}_{2}\Lambda^{3}{}_{3}F^{23} = 1\times 1\times B_x \\ &= B_x, \\ B_{y'} &= F^{3'1'} = \Lambda^{3}{}_{\mu}\Lambda^{1}{}_{\nu}F^{\mu\nu} = \Lambda^{3}{}_{3}\Lambda^{1}{}_{\nu}F^{3\nu} = \Lambda^{3}{}_{3}\Lambda^{1}{}_{0}F^{30} + \Lambda^{3}{}_{3}\Lambda^{1}{}_{1}F^{31} \\ &= 1\times(-\beta\gamma)(-E_z) + 1\times\gamma B_y = \gamma B_y + \beta\gamma E_z \\ &= \gamma\left(B - \beta\times E\right)_y \end{aligned}$$

$$\begin{aligned}
B_{z'} &= F^{1'2'} = \Lambda^{1}{}_{\mu}\Lambda^{2}{}_{\nu}F^{\mu\nu} = \Lambda^{1}{}_{\mu}\Lambda^{2}{}_{2}F^{\mu 2} = \Lambda^{1}{}_{0}\Lambda^{2}{}_{2}F^{02} + \Lambda^{1}{}_{1}\Lambda^{2}{}_{2}F^{12} \\
&= (-\gamma\beta)\times 1\times E_y + \gamma\times 1\times B_z = \gamma B_z - \beta\gamma E_y \\
&= \gamma\left(B - \beta\times E\right)_z
\end{aligned}$$

For the electric field results,

$$\begin{aligned}
E_{x'} &= F^{0'1'} = \Lambda^{0}{}_{\mu}\Lambda^{1}{}_{\nu}F^{\mu\nu} = \Lambda^{0}{}_{1}\Lambda^{1}{}_{0}F^{10} + \Lambda^{0}{}_{0}\Lambda^{1}{}_{1}F^{01} \\
&= (-\gamma\beta)(-\gamma\beta)(-E_x) + \gamma\gamma E_x = -\gamma^2\beta^2(E_x) + \gamma^2 E_x = E_x(1-\beta^2)\gamma^2 \\
&= E_x, \\
E_{y'} &= F^{0'2'} = \Lambda^{0}{}_{\mu}\Lambda^{2}{}_{\nu}F^{\mu\nu} = \Lambda^{0}{}_{\mu}\Lambda^{2}{}_{2}F^{\mu 2} = \Lambda^{0}{}_{0}\Lambda^{2}{}_{2}F^{02} + \Lambda^{0}{}_{1}\Lambda^{2}{}_{2}F^{12} \\
&= \gamma\times 1\times E_y + (-\beta\gamma)\times 1\times B_z = \gamma E_y - \beta\gamma B_z \\
&= \gamma\left(E + \beta\times B\right)_y \\
E_{z'} &= F^{0'3'} = \Lambda^{0}{}_{\mu}\Lambda^{3}{}_{\nu}F^{\mu\nu} = \Lambda^{0}{}_{\mu}\Lambda^{3}{}_{3}F^{\mu 3} = \Lambda^{0}{}_{0}\Lambda^{3}{}_{3}F^{03} + \Lambda^{0}{}_{1}\Lambda^{3}{}_{3}F^{13} \\
&= \gamma\times 1\times E_z - \beta\gamma\times 1\times(-B_y) = \gamma E_z + \beta\gamma B_y \\
&= \gamma\left(E + \beta\times B\right)_z .
\end{aligned}$$

Here, $\beta = (\beta, 0, 0)$ is used. These results can be summarized by,

$$\begin{aligned}
E_{\|'} &= E_{\|} \\
B_{\|'} &= B_{\|} \\
E_{\perp'} &= \gamma\left(E_{\perp} + \beta\times B_{\perp}\right) = \gamma\left(E + \beta\times B\right)_{\perp}, \\
B_{\perp'} &= \gamma\left(B_{\perp} - \beta\times E_{\perp}\right) = \gamma\left(B - \beta\times E\right)_{\perp},
\end{aligned}$$

and are independent of the metric signature. For SI units, substitute $E \rightarrow {}^{E}\!/_{c}$. Misner, Thorne & Wheeler (1973) refer to this last form as the 3 + 1 view as opposed to the *geometric view* represented by the tensor expression,

$$F^{\mu'\nu'} = \Lambda^{\mu'}{}_{\mu}\Lambda^{\nu'}{}_{\nu}F^{\mu\nu},$$

and make a strong point of the ease with which results that are difficult to achieve using the 3 + 1 view can be obtained and understood. Only objects that have well defined Lorentz transformation properties (in fact under *any* smooth coordinate transformation) are geometric objects. In the geometric view, the electromagnetic field is a six-dimensional geometric object in *spacetime* as opposed to two interdependent, but separate, 3-vector fields in *space* and *time*. The fields E (alone) and B (alone) do not have well defined Lorentz transformation properties. The mathematical underpinnings are two equations mentioned in the starting of the topic that immediately yield the third equation mentioned in the starting of the topic. One should note that the primed and unprimed tensors refer to the *same event in spacetime*. Thus the complete equation with spacetime dependence is,

$$F^{\mu'\nu'}\left(x'\right) = \Lambda^{\mu'}{}_{\mu}\Lambda^{\nu'}{}_{\nu}F^{\mu\nu}\left(\Lambda^{-1}x'\right) = \Lambda^{\mu'}{}_{\mu}\Lambda^{\nu'}{}_{\nu}F^{\mu\nu}(x).$$

Length contraction has an effect on charge density ρ and current density J, and time dilation has an effect on the rate of flow of charge (current), so charge and current distributions must transform in a related way under a boost. It turns out they transform exactly like the space-time and energy-momentum four-vectors,

$$\mathrm{j}' = \mathrm{j} - \gamma\rho v\mathrm{n} + (\gamma - 1)(\mathrm{j}\cdot\mathrm{n})\mathrm{n}$$

$$\rho' = \gamma\left(\rho - \mathrm{j}\cdot\frac{v\mathrm{n}}{\mathrm{c}^2}\right),$$

or, in the simpler geometric view,

$$j^{\mu'} = \Lambda^{\mu'}{}_{\mu} j^{\mu}.$$

One says that charge density transforms as the time component of a four-vector. It is a rotational scalar. The current density is a 3-vector.

The Maxwell equations are invariant under Lorentz transformations.

Spinors

Equation $(A\otimes B)(u\otimes v) = Au\otimes Bv, \quad u\in U, v\in V, u\otimes v\in U\otimes V.$ hold unmodified for any representation of the Lorentz group, including the bispinor representation. In $u\otimes v \to \Lambda u\otimes \Lambda v = \Lambda^{\mu}{}_{\nu}u^{\nu}\otimes\Lambda^{\rho}{}_{\sigma}v^{\sigma} = \Lambda^{\mu}{}_{\nu}\Lambda^{\rho}{}_{\sigma}u^{\nu}\otimes v^{\sigma} \equiv \Lambda^{\mu}{}_{\nu}\Lambda^{\rho}{}_{\sigma}w^{\nu\sigma}$. one simply replaces all occurrences of Λ by the bispinor representation Π(Λ),

$$u\otimes v \to \Pi(\Lambda)u\otimes\Pi(\Lambda)v = \Pi(\Lambda)^{\alpha}{}_{\beta}u^{\beta}\otimes\Pi(\Lambda)^{\rho}{}_{\sigma}v^{\sigma} = \Pi(\Lambda)^{\alpha}{}_{\beta}\Pi(\Lambda)^{\rho}{}_{\sigma}u^{\beta}\otimes v^{\sigma} \equiv \Pi(\Lambda)^{\alpha}{}_{\beta}\Pi(\Lambda)^{\rho}{}_{\sigma}w^{\alpha\beta}.$$

The above equation could, for instance, be the transformation of a state in Fock space describing two free electrons.

Transformation of General Fields

A general *noninteracting* multi-particle state (Fock space state) in quantum field theory transforms according to the rule,

$$U(\Lambda,a)\Psi_{p_1\sigma_1 n_1; p_2\sigma_2 n_2;\cdots}$$

$$= e^{-ia_\mu\left[(\Lambda p_1)^\mu + (\Lambda p_2)^\mu + \cdots\right]}\sqrt{\frac{(\Lambda p_1)^0(\Lambda p_2)^0\cdots}{p_1^0 p_2^0\cdots}}\left(\sum_{\sigma_1'\sigma_2'\cdots} D^{(j_1)}_{\sigma_1'\sigma_1}\left[W(\Lambda,p_1)\right]D^{(j_2)}_{\sigma_2'\sigma_2}\left[W(\Lambda,p_2)\right]\cdots\right)\Psi_{\Lambda p_1\sigma_1' n_1;\Lambda p_2\sigma_2' n_2;\cdots},$$

where $W(\Lambda, p)$ is the Wigner rotation and $D^{(j)}$ is the $(2j + 1)$-dimensional representation of SO(3).

Time Dilation

Time dilation, in the theory of special relativity, the "slowing down" of a clock as determined by an observer who is in relative motion with respect to that clock. In special relativity, an observer in inertial(i.e., nonaccelerating) motion has a well-defined means of determining which events occur simultaneously with a given event. A second inertial observer, who is in relative motion with respect to the first, however, will disagree with the first observer regarding which events are simultaneous with that given event. (Neither observer is wrong in this determination; rather, their disagreement merely reflects the fact that simultaneity is an observer-dependent notion in special relativity.) A notion of simultaneity is required in order to make a comparison of the rates of clocks carried by the two observers. If the first observer's notion of simultaneity is used, it is found that the secondobserver's clock runs slower than the first observer's by a factor of Square root of $\sqrt{(1-v^2/c^2)}$, where *v* is the relative velocity of the observers and *c* equals 299, 792 km (186, 282 miles) per second—i.e., the speed of light. Similarly, using the second observer's notion of simultaneity, it is found that the first observer's clock runs slower by the same factor. Thus, each inertial observer determines that all clocks in motion relative to that observer run slower than that observer's own clock.

A closely related phenomenon predicted by special relativity is the so-called twin paradox. Suppose one of two twins carrying a clock departs on a rocket ship from the other twin, an inertial observer, at a certain time, and they rejoin at a later time. In accordance with the time-dilation effect, the elapsed time on the clock of the twin on the rocket ship will be smaller than that of the inertial observer twin—i.e., the non-inertial twin will have aged less than the inertial observer twin when they rejoin.

The time-dilation effect predicted by special relativity has been accurately confirmed by observations of the increased lifetime of unstable elementary particles traveling at nearly the speed of light. The clock paradox effect also has been substantiated by experiments comparing the elapsed time of an atomic clock on Earth with that of an atomic clock flown in an airplane. The latter experiments, furthermore, have confirmed a gravitational contribution to time dilation, as predicted by the theory of general relativity.

The equation for calculating time dilation is as follows:

$t = t_0/(1\text{-}v^2/c^2)^{1/2}$

where: t = time observed in the other reference frame

t_0 = time in observers own frame of reference (rest time)

v = the speed of the moving object

c = the speed of light in a vacuum

so in our problem we will let v = .95c, t_0 = 10 years and we will solve for t which is the time that the earth bound brother measures.

$t = 10/(1\text{-}\ (.95c)^2/c^2)^{1/2}$

$t = 10/(1\text{-}\ .95^2)^{1/2}$

$t = 10/\ .312$

t = 32 years (the time the earth bound brother measures)

Now let's have a closer look at the equation and determine just what impact the speed of the object has on time dilation. We can see that is the velocity is small compared to the speed of light the quantity v^2/c^2 approaches 0 and the equation simplifies t_0: $t = t_0/1$ which is simply t. So at relatively slow speeds (our everyday speeds) time dilation is not a factor and Newton's Laws are still applicable. Now let's look at high speeds (close to the speed of light), from the equation that as velocity increases the quantity v^2/c^2approaches 1 (but will never quit reach it), causing the quantity$(1-v^2/c^2)^{1/2}$ t_0 become smaller and smaller....therefore causing the time measured by the other observer t_0 become greater thus making our time appear slower (refer back to the example). I know its so confusing!!! read it again, think about it, then study the graph below. As one can see in the graph time dilation starts t_0 "show up" between .4c and .5c. Also notice that the closer one gets to the speed of light the greater impact speed has on time dilation (notice how steep the curve gets towards the end).

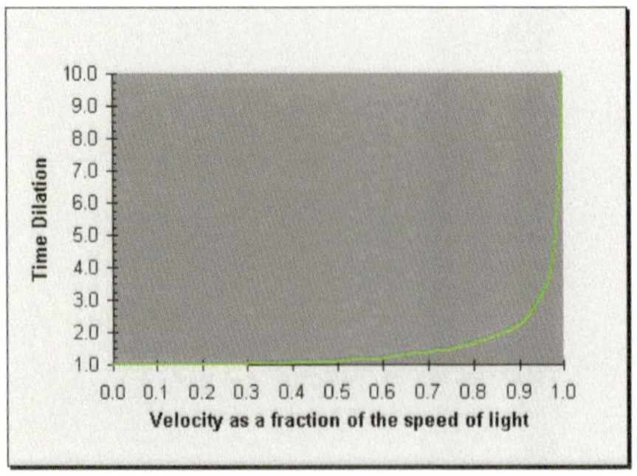

Velocity Time Dilation

Special relativity indicates that, for an observer in an inertial frame of reference, a clock that is moving relative to him will be measured to tick slower than a clock that is at rest in his frame of reference. This case is sometimes called special relativistic time dilation. The faster the relative velocity, the greater the time dilation between one another, with the rate of time reaching zero as one approaches the speed of light (299, 792, 458 m/s). This causes massless particles that travel at the speed of light to be unaffected by the passage of time.

Theoretically, time dilation would make it possible for passengers in a fast-moving vehicle to advance further into the future in a short period of their own time. For sufficiently high speeds, the effect is dramatic. For example, one year of travel might correspond to ten years on Earth. Indeed, a constant 1 g acceleration would permit humans to travel through the entire known Universe in one human lifetime.. At a constant 1 g traveling up to 0.99999999 c it would take 30 years to reach the edge of the universe 13.5 billions lightyears away. Space travelers could then return to Earth billions of years in the future. A scenario based on this idea was presented in the novel *Planet of the Apes* by Pierre Boulle, and the Orion Project has been an attempt toward this idea.

With current technology severely limiting the velocity of space travel, however, the differences experienced in practice are minuscule: after 6 months on the International Space Station (ISS) (which orbits Earth at a speed of about 7, 700 m/s) an astronaut would have aged about 0.005 seconds less than those on Earth. The current human time travel record holder is Russian cosmonaut Sergei Krikalev. He gained 22.68 milliseconds of lifetime on his journeys to space and therefore beat the previous record of about 20 milliseconds by cosmonaut Sergei Avdeyev.

Simple Inference of Velocity Time Dilation

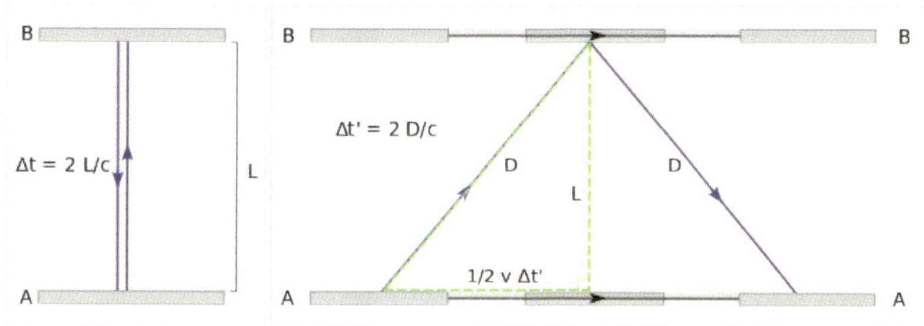

Left: Observer at rest measures time $2L/c$ between co-local events of light signal generation at A and arrival at A. Right: Events according to an observer moving to the left of the setup: bottom mirror A when signal is generated at time $t'=0$, top mirror B when signal gets reflected at time $t'=D/c$, bottom mirror A when signal returns at time $t'=2D/c$

Time dilation can be inferred from the observed constancy of the speed of light in all reference frames dictated by the second postulate of special relativity.

This constancy of the speed of light means that, counter to intuition, speeds of material objects and light are not additive. It is not possible to make the speed of light appear greater by moving towards or away from the light source.

Consider then, a simple clock consisting of two mirrors A and B, between which a light pulse is bouncing. The separation of the mirrors is L and the clock ticks once each time the light pulse hits either of the mirrors.

In the frame in which the clock is at rest (diagram on the left), the light pulse traces out a path of length $2L$ and the period of the clock is $2L$divided by the speed of light:

$$\Delta t = \frac{2L}{c}.$$

From the frame of reference of a moving observer traveling at the speed v relative to the resting frame of the clock (diagram at right), the light pulse is seen as tracing out a longer, angled path. Keeping the speed of light constant for all inertial observers, requires a lengthening of the period of this clock from the moving observer's perspective. That is to say, in a frame moving relative to the local clock, this clock will appear to be running more slowly. Straightforward application of the Pythagorean theorem leads to the well-known prediction of special relativity:

The total time for the light pulse to trace its path is given by,

$$\Delta t' = \frac{2D}{c}.$$

The length of the half path can be calculated as a function of known quantities as,

$$D = \sqrt{\left(\frac{1}{2}v\Delta t'\right)^2 + L^2}.$$

Elimination of the variables D and L from these three equations results in,

$$\Delta t' = \frac{\Delta t}{\sqrt{1 - \frac{v^2}{c^2}}},$$

which expresses the fact that the moving observer's period of the clock Ät′ is longer than the period Δt in the frame of the clock itself.

Reciprocity

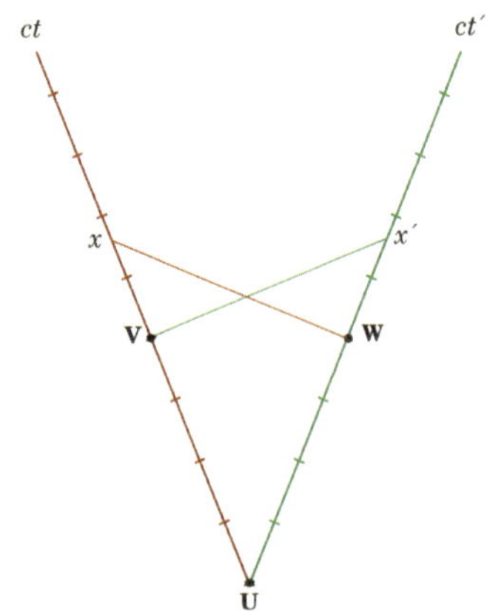

Time UV of a clock in S is shorter compared to Ux′ in S′, and time UW of a clock in S′ is shorter compared to Ux in S

Given a certain frame of reference, and the "stationary" observer, if a second observer accompanied the "moving" clock, each of the observers would perceive the other's clock as ticking at a *slower* rate than their own local clock, due to them both perceiving the other to be the one that's in motion relative to their own stationary frame of reference.

Common sense would dictate that, if the passage of time has slowed for a moving object, said object would observe the external world's time to be correspondingly sped up. Counterintuitively, special relativity predicts the opposite. When two observers are in motion relative to each other, each will measure the other's clock slowing down, in concordance with them being moving relative to the observer's frame of reference.

While this seems self-contradictory, a similar oddity occurs in everyday life. If two persons A and B observe each other from a distance, B will appear small to A, but at the same time A will appear small to B. Being familiar with the effects of perspective, there is no contradiction or paradox in this situation.

The reciprocity of the phenomenon also leads to the so-called twin paradox where the aging of twins, one staying on Earth and the other embarking on a space travel, is compared, and where the reciprocity suggests that both persons should have the same age when they reunite. On the contrary, at the end of the round-trip, the traveling twin will be younger than his brother on Earth. The dilemma posed by the paradox, however, can be explained by the fact that the traveling twin must markedly accelerate in at least three phases of the trip (beginning, direction change, and end), while the other will only experience negligible acceleration, due to rotation and revolution of Earth. During the acceleration phases of the space travel, time dilation is not symmetric.

Experimental Testing

Doppler Effect

- Ives and Stilwell (1938, 1941) stated purpose of these experiments was to verify the time dilation effect, predicted by Larmor–Lorentz ether theory, due to motion through the ether using Einstein's suggestion that Doppler effect in canal rays would provide a suitable experiment. These experiments measured the Doppler shift of the radiation emitted from cathode rays, when viewed from directly in front and from directly behind. The high and low frequencies detected were not the classically predicted values,

$$\frac{f_0}{1-v/c} \quad \text{and} \quad \frac{f_0}{1+v/c}.$$

The high and low frequencies of the radiation from the moving sources were measured as,

$$\sqrt{\frac{1+v/c}{1-v/c}}f_0=\gamma(1+v/c)f_0 \quad \text{and} \quad \sqrt{\frac{1-v/c}{1+v/c}}f_0=\gamma(1-v/c)f_0,$$

as deduced by Einstein (1905) from the Lorentz transformation, when the source is running slow by the Lorentz factor.

- Hasselkamp, Mondry, and Scharmann (1979) measured the Doppler shift from a source moving at right angles to the line of sight. The most general relationship between frequencies of the radiation from the moving sources is given by:

$$f_{\text{detected}}=f_{\text{rest}}\left(1-\frac{v}{c}\cos\phi\right)/\sqrt{1-v^2/c^2}$$

as deduced by Einstein (1905). For ϕ = 90° (cos ϕ = 0) this reduces to fdetected = frestγ. This lower frequency from the moving source can be attributed to the time dilation effect and is often called the transverse Doppler effect and was predicted by relativity.

- In 2010 time dilation was observed at speeds of less than 10 meters per second using optical atomic clocks connected by 75 meters of optical fiber.

Moving Particles

- A comparison of muon lifetimes at different speeds is possible. In the laboratory, slow muons are produced; and in the atmosphere, very fast moving muons are introduced by cosmic rays. Taking the muon lifetime at rest as the laboratory value of 2.197 μs, the lifetime of a cosmic ray produced muon traveling at 98% of the speed of light is about five times longer, in agreement with observations. An example is Rossi and Hall (1941), who compared the population of cosmic-ray-produced muons at the top of a mountain to that observed at sea level.

- The lifetime of particles produced in particle accelerators appears longer due to time dilation. In such experiments the "clock" is the time taken by processes leading to muon decay, and these processes take place in the moving muon at its own "clock rate", which is much slower than the laboratory clock. This is routinely taken into account in particle physics, and many dedicated measurements have been performed. For instance, in the muon storage ring at CERN the lifetime of muons circulating with $\gamma = 29.327$ was found to be dilated to 64.378 μs, confirming time dilation to an accuracy of 0.9 ± 0.4 parts per thousand.

Proper Time and Minkowski Diagram

Minkowski Diagram and Twin Paradox

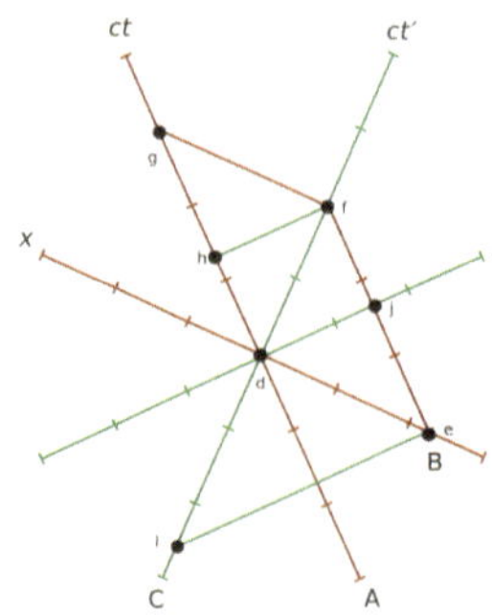

Clock C in relative motion between two synchronized clocks A and B. C meets A at *d*, and B at *f*

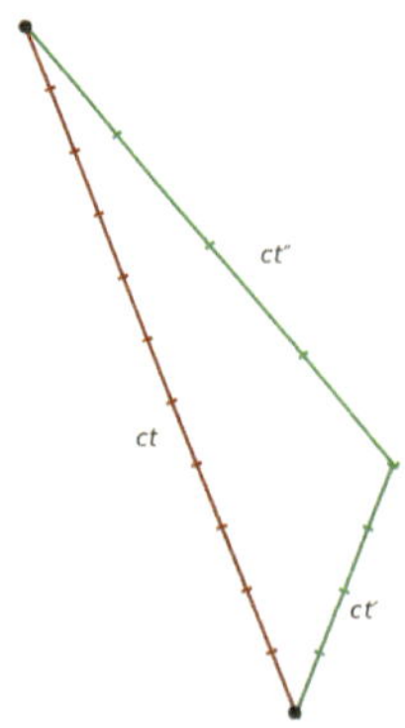

Twin paradox. One twin has to change frames, leading to different proper times in the twin's world lines

In the Minkowski diagram from the second image on the right, clock C resting in inertial frame S′ meets clock A at *d* and clock B at *f* (both resting in S). All three clocks simultaneously start to tick in S. The worldline of A is the ct-axis, the worldline of B intersecting *f* is parallel to the ct-axis, and the worldline of C is the ct′-axis. All events simultaneous with *d* in S are on the x-axis, in S′ on the x′-axis.

The proper time between two events is indicated by a clock present at both events. It is invariant, i.e., in all inertial frames it is agreed that this time is indicated by that clock. Interval *df* is therefore the proper time of clock C, and is shorter with respect to the coordinate times *ef=dg* of clocks B and A in S. Conversely, also proper time *ef* of B is shorter with respect to time *if* in S′, because event *e* was measured in S′ already at time *i*due to relativity of simultaneity, long before C started to tick.

From that it can be seen, that the proper time between two events indicated by an unaccelerated clock present at both events, compared with the synchronized coordinate time measured in all other inertial frames, is always the *minimal* time interval between those events. However, the interval between two events can also correspond to the proper time of accelerated clocks present at both events. Under all possible proper times between two events, the proper time of the unaccelerated clock is *maximal*, which is the solution to the twin paradox.

Derivation and Formulation

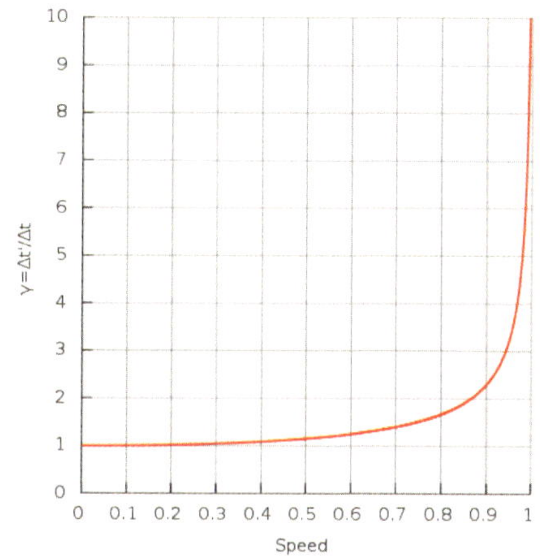

Lorentz factor as a function of speed (in natural units where $c = 1$).
Notice that for small speeds (less than 0.1), γ is approximately 1

In addition to the light clock used above, the formula for time dilation can be more generally derived from the temporal part of the Lorentz transformation. Let there be two events at which the moving clock indicates t_a and t_b, thus

$$t_a' = \frac{t_a - \frac{vx_a}{c^2}}{\sqrt{1-\frac{v^2}{c^2}}}, t_b' = \frac{t_b - \frac{vx_b}{c^2}}{\sqrt{1-\frac{v^2}{c^2}}}.$$

Since the clock remains at rest in its inertial frame, it follows $x_a = x_b$, thus the interval $\Delta t' = t_b' - t_a'$ is given by,

$$\Delta t' = \gamma \Delta t = \frac{\Delta t}{\sqrt{1-\frac{v^2}{c^2}}}$$

where Δt is the time interval between *two co-local events* (i.e. happening at the same place) for an observer in some inertial frame (e.g. ticks on his clock), known as the *proper time*, $\Delta t'$ is the time interval between those same events, as measured by another observer, inertially moving with velocity v with respect to the former observer, v is the relative velocity between the observer and the moving clock, c is the speed of light, and the Lorentz factor (conventionally denoted by the Greek letter gamma or γ) is,

$$\gamma = \frac{1}{\sqrt{1-\frac{v^2}{c^2}}}.$$

Thus the duration of the clock cycle of a moving clock is found to be increased: it is measured to be "running slow". The range of such variances in ordinary life, where $v \ll c$, even considering space travel, are not great enough to produce easily detectable time dilation effects and such vanishingly small effects can be safely ignored for most purposes. It is only when an object approaches speeds on the order of 30, 000 km/s (1/10 the speed of light) that time dilation becomes important.

Hyperbolic Motion

In special relativity, time dilation is most simply described in circumstances where relative velocity is unchanging. Nevertheless, the Lorentz equations allow one to calculate proper time and movement in space for the simple case of a spaceship which is applied with a force per unit mass, relative to some reference object in uniform (i.e. constant velocity) motion, equal to g throughout the period of measurement.

Let t be the time in an inertial frame subsequently called the rest frame. Let x be a spatial coordinate, and let the direction of the constant acceleration as well as the spaceship's velocity (relative to the rest frame) be parallel to the x-axis. Assuming the spaceship's position at time $t = 0$ being $x = 0$ and the velocity being v_0 and defining the following abbreviation,

$$\gamma_0 = \frac{1}{\sqrt{1 - v_0^2 / c^2}},$$

the following formulas hold:

Position:

$$x(t) = \frac{c^2}{g}\left(\sqrt{1 + \frac{(gt + v_0\gamma_0)^2}{c^2}} - \gamma_0\right).$$

Velocity:

$$v(t) = \frac{gt + v_0\gamma_0}{\sqrt{1 + \frac{(gt + v_0\gamma_0)^2}{c^2}}}.$$

Proper time as function of coordinate time:

$$\tau(t) = \tau_0 + \int_0^t \sqrt{1 - \left(\frac{v(t')}{c}\right)^2}\, dt'.$$

In the case where $v(0) = v_0 = 0$ and $\tau(0) = \tau_0 = 0$ the integral can be expressed as a logarithmic function or, equivalently, as an inverse hyperbolic function:

$$\tau(t) = \frac{c}{g} \ln\left(\frac{gt}{c} + \sqrt{1 + \left(\frac{gt}{c}\right)^2}\right) = \frac{c}{g} \operatorname{arsinh}\left(\frac{gt}{c}\right).$$

As functions of the proper time τ of the ship, the following formulae hold:

Position:

$$x(\tau)=\frac{c^2}{g}\left(cosh\frac{g\tau}{c}-1\right).$$

Velocity:

$$v(\tau)=c\,tanh\frac{g\tau}{c}.$$

Coordinate time as function of proper time:

$$t(\tau)=\frac{c}{g}sinh\frac{g\tau}{c}.$$

Clock Hypothesis

The clock hypothesis is the assumption that the rate at which a clock is affected by time dilation does not depend on its acceleration but only on its instantaneous velocity. This is equivalent to stating that a clock moving along a path P measures the proper time, defined by:

$$d\tau=\int_P\sqrt{dt^2-dx^2/c^2-dy^2/c^2-dz^2/c^2}.$$

The clock hypothesis was implicitly (but not explicitly) included in Einstein's original 1905 formulation of special relativity. Since then, it has become a standard assumption and is usually included in the axioms of special relativity, especially in the light of experimental verification up to very high accelerations in particle accelerators.

Gravitational time dilation is a phenomenon whereby time runs slower when in a higher gravitational potential. Put simply, the closer you are to a large body like the Earth the slower time runs, thus time runs slower for someone on the surface of the earth compared to someone in orbit around the earth.

Whilst the Einstein's field equation for General relativity is very complicated the equation for time dilation is much simpler and bears certain resemblance to the equation for time dilation for relative motion. The equation is:

$$t_o=t_f\sqrt{1-\frac{2GM}{rc^2}}$$

In this equation to is the proper time for an event, i.e. the time measured when observer and event are in the same gravitational potential and tf is the time as measured when at an infinite distance from any mass. The values G and c are again the Newtonian gravitational potential and speed of light respectively and M and r are the mass of the object you are near and r is your distance from said object respectively.

This equation can be written differently by applying Schwarzschild's solution to general relativity which gives you the Schwarzschild radius; the radius of an object of mass M, for which if all mass is contained within that radius then the gravitational field is too strong that not even light can escape.

$$r_s = \frac{2GM}{c^2} \Rightarrow t_o = t_f\sqrt{1-\frac{2GM}{rc^2}} = t_f\sqrt{1-\frac{r_s}{r}}$$

If one therefore applies this formula whose mass is contained within their Schwarzschild radius, i.e. Black Holes, then when something or someone reaches the edge of a black hole when r=rs time for the object/person appears to stop from the point of view of an observer further away. Thus you can never see something fall into a black hole as time stops at the edge of the black hole.

It then gets even stranger once the person has fallen into the black hole (although you will never see it) as the value in the square root operator becomes negative so the answer come back as an imaginary number and no one knows how to interpret an imaginary result to the real world.

References

- Bartusiak, Marcia (2000), Einstein's Unfinished Symphony: Listening to the Sounds of Space-Time, Berkley, ISBN 978-0-425-18620-6
- Relativistic-mechanics, science: britannica.com, Retrieved 29 June 2018
- Bekenstein, Jacob D. (1973), «Black Holes and Entropy», Phys. Rev., D7 (8): 2333–2346, Bibcode:1973PhRvD...7.2333B, doi:10.1103/PhysRevD.7.2333
- Theory-of-relativity: allaboutscience.org, Retrieved 09 July 2018
- Blanchet, Luc (2006), "Gravitational Radiation from Post-Newtonian Sources and Inspiralling Compact Binaries", Living Reviews in Relativity, 9, Bibcode:2006LRR.....9....4B, doi:10.12942/lrr-2006-4, retrieved 2007-08-07
- General-relativity: thestargarden.co.uk, Retrieved 29 June 2018
- Blandford, R. D. (1987), «Astrophysical Black Holes», in Hawking, Stephen W.; Israel, Werner, 300 Years of Gravitation, Cambridge University Press, pp. 277–329, ISBN 0-521-37976-8
- Physics, relativistic-mass-formula-546: softschools.com, Retrieved 11 April 2018
- What-is-relativistic-mass-and-why-it-is-not-used-much: physicsforums.com, Retrieved 18 May 2018
- Bekenstein, Jacob D. (1974), "Generalized Second Law of Thermodynamics in Black-Hole Physics", Phys. Rev., D9 (12): 3292–3300, Bibcode:1974PhRvD...9.3292B, doi:10.1103/PhysRevD.9.3292
- Derivation-of-relativistic-mass: thecuriousastronomer.wordpress.com, Retrieved 22 July 2018
- Physics, relativistic-energy-formula-545: softschools.com, Retrieved 28 June 2018
- Font, José A. (2003), "Numerical Hydrodynamics in General Relativity", Living Reviews in Relativity, 6, Bibcode:2003LRR.....6....4F, doi:10.12942/lrr-2003-4, PMC 5256036, PMID 28179854, retrieved 2007-08-19
- Lorentz-transformations, science: britannica.com, Retrieved 08 July 2018
- Time-dilation, science: britannica.com, Retrieved 27 July 2018
- What-is-gravitational-time-dilation: users.sussex.ac.uk, Retrieved 18 April 2018

4

Statistical Mechanics

In physics, statistical mechanics is the approach to the study of a system with many degrees of freedom. This approach is based on an understanding of probability theory, statistical methods and physical laws. In order to completely understand statistical mechanics, it is necessary to study the concepts of microstate and macrostate, statistical ensemble, partition function, etc. which have been extensively discussed in this chapter.

Statistical mechanics is either the least fundamental or most fundamental of all fields of physics. That is because it is not really science at all. It is pure mathematics.

In other subjects, you learn about "natural laws": Newton's second law, Maxwell's equations, Schrödinger's equation, etc. These laws are not derived from anything else. They were discovered experimentally and then assumed to reflect fundamental aspects of reality. But statistical mechanics does not involve any natural laws of this sort. Instead, it is a set of techniques that can be applied to nearly any physical system, no matter what laws that system obeys. That is why I call it the most fundamental field of physics. New theories may replace old ones, and natural laws may turn out to be merely approximations to deeper laws. But statistical mechanics remains valid through it all, and whatever new laws are discovered, it will almost certainly work just as well with them as it did with the old ones.

Statistical mechanics was developed in the second half of the 19th century. It was primarily the work of Ludwig Boltzmann, who personally published more than a hundred papers on the subject during his lifetime. Other scientists also contributed to it, of course, especially James Clerk Maxwell (the same one Maxwell's equations are named after) and Josiah Willard Gibbs.

It grew out of thermodynamics, an earlier theory that described the behavior of a mysterious substance called "heat" or "caloric". Thermodynamics was a physical theory of the more conventional sort. It involved natural laws discovered by experiment, and made no claims about why those laws happened to hold. Caloric was supposed to be a substance much like other forms of matter. But this view turned out to be incorrect. Heat is actually an emergent phenomenon: a mathematical quantity defined in terms of a more detailed theory (the movement of individual atoms). The "laws" of thermodynamics can be derived from that deeper theory by applying statistical techniques. If you ignore the details of how each atom is moving, you find they collectively behave in a way that resembles a continuous fluid. That is what statistical mechanics is all about: deriving high level descriptions by starting from lower level ones and then averaging out lots of details.

Is it possible other theories could be explained in the same way? That is an open question, and a fascinating one. Many physicists suspect gravity is an emergent phenomenon, that it arises from the collective behavior of some deeper degrees of freedom. Statistical interpretations have also been proposed for quantum mechanics. These are all very speculative, of course, and they could easily turn out to be wrong. But they also could easily turn out to be right. Based on what we know today, it is entirely possible that the very structure of spacetime is a consequence of statistical mechanics.

Fundamentals of Statistical Mechanics

Statistical mechanics consists of the study of the particular laws which govern the behavior and properties of macroscopic bodies, that is to say, bodies made up of a very large number of separate particles (atoms and molecules). The number of particles is of the order of 10^{23}, the Avogadro number. From microscopic point of view, complete information about the system can be obtained by setting up the equations of motion and then solving them. In general, equations of motion are second order differential equations in both classical and quantum case (and twice the number of first order differential equations in Hamiltonian formulation). Their solutions will have two constants of integration for each equation of motion, corresponding to each particle, to be determined from the initial conditions of individual particles. This is a tremendous task involving a lot of paperwork and is a near impossibility. But a new type of regularity can be found by relating the macroscopic description to a few variables based on statistical laws, which arise due to the large number of particles involved. These laws become quite meaningless if applied to a single particle or even a mechanical system with very few degrees of freedom. The development and exposition of these laws constitute the subject matter of statistical mechanics, which arise due to the large number of particles involved.

The importance of statistical mechanics lies in the fact that, in nature, we are dealing all the time with macroscopic bodies (matter in a bulk) whose behavior cannot be described by pure mechanical methods and which in fact do obey statistical laws.

Principles: Mechanics and Ensembles

In physics, there are two types of mechanics usually examined: classical mechanics and quantum mechanics. For both types of mechanics, the standard mathematical approach is to consider two concepts:

1. The complete state of the mechanical system at a given time, mathematically encoded as a phase point (classical mechanics) or a pure quantum state vector (quantum mechanics).
2. An equation of motion which carries the state forward in time: Hamilton's equations (classical mechanics) or the time-dependent Schrödinger equation(quantum mechanics).

Using these two concepts, the state at any other time, past or future, can in principle be calculated. There is however a disconnection between these laws and everyday life experiences, as we do not find it necessary (nor even theoretically possible) to know exactly at a microscopic level the simultaneous positions and velocities of each molecule while carrying out processes at the human scale (for example, when performing a chemical reaction). Statistical mechanics fills this disconnection between the laws of mechanics and the practical experience of incomplete knowledge, by adding some uncertainty about which state the system is in.

Whereas ordinary mechanics only considers the behaviour of a single state, statistical mechanics introduces the statistical ensemble, which is a large collection of virtual, independent copies of the system in various states. The statistical ensemble is a probability distribution over all possible states of the system. In classical statistical mechanics, the ensemble is a probability distribution over phase points (as opposed to a single phase point in ordinary mechanics), usually represented as a distribution in a phase space with canonical coordinates. In quantum statistical mechanics, the ensemble is a probability distribution over pure states, and can be compactly summarized as a density matrix.

As is usual for probabilities, the ensemble can be interpreted in different ways:

- An ensemble can be taken to represent the various possible states that a *single system* could be in (epistemic probability, a form of knowledge), or
- The members of the ensemble can be understood as the states of the systems in experiments repeated on independent systems which have been prepared in a similar but imperfectly controlled manner (empirical probability), in the limit of an infinite number of trials.

These two meanings are equivalent for many purposes.

However the probability is interpreted, each state in the ensemble evolves over time according to the equation of motion. Thus, the ensemble itself (the probability distribution over states) also evolves, as the virtual systems in the ensemble continually leave one state and enter another. The ensemble evolution is given by the Liouville equation (classical mechanics) or the von Neumann equation (quantum mechanics). These equations are simply derived by the application of the mechanical equation of motion separately to each virtual system contained in the ensemble, with the probability of the virtual system being conserved over time as it evolves from state to state.

One special class of ensemble is those ensembles that do not evolve over time. These ensembles are known as *equilibrium ensembles* and their condition is known as *statistical equilibrium*. Statistical equilibrium occurs if, for each state in the ensemble, the ensemble also contains all of its future and past states with probabilities equal to the probability of being in that state. The study of equilibrium ensembles of isolated systems is the focus of statistical thermodynamics. Non-equilibrium statistical mechanics addresses the more general case of ensembles that change over time, and/or ensembles of non-isolated systems.

Statistical Thermodynamics

The primary goal of statistical thermodynamics (also known as equilibrium statistical mechanics) is to derive the classical thermodynamics of materials in terms of the properties of their constituent particles and the interactions between them. In other words, statistical thermodynamics provides a connection between the macroscopic properties of materials in thermodynamic equilibrium, and the microscopic behaviours and motions occurring inside the material.

Whereas statistical mechanics proper involves dynamics, here the attention is focussed on *statistical equilibrium* (steady state). Statistical equilibrium does not mean that the particles have stopped moving (mechanical equilibrium), rather, only that the ensemble is not evolving.

Fundamental Postulate

A sufficient (but not necessary) condition for statistical equilibrium with an isolated system is that the probability distribution is a function only of conserved properties (total energy, total particle numbers, etc.). There are many different equilibrium ensembles that can be considered, and only some of them correspond to thermodynamics. Additional postulates are necessary to motivate why the ensemble for a given system should have one form or another.

A common approach found in many textbooks is to take the *equal a priori probability postulate.* This postulate states that,

> *For an isolated system with an exactly known energy and exactly known composition, the system can be found with* equal probability *in any microstate consistent with that knowledge.*

The equal a priori probability postulate therefore provides a motivation for the microcanonical ensemble. There are various arguments in favour of the equal a priori probability postulate:

- Ergodic hypothesis: An ergodic system is one that evolves over time to explore "all accessible" states: all those with the same energy and composition. In an ergodic system, the microcanonical ensemble is the only possible equilibrium ensemble with fixed energy. This approach has limited applicability, since most systems are not ergodic.
- Principle of indifference: In the absence of any further information, we can only assign equal probabilities to each compatible situation.
- Maximum information entropy: A more elaborate version of the principle of indifference states that the correct ensemble is the ensemble that is compatible with the known information and that has the largest Gibbs entropy (information entropy).

Other fundamental postulates for statistical mechanics have also been proposed.

Three Thermodynamic Ensembles

There are three equilibrium ensembles with a simple form that can be defined for any isolated system bounded inside a finite volume. These are the most often discussed ensembles in statistical thermodynamics. In the macroscopic limit they all correspond to classical thermodynamics.

- Microcanonical ensemble describes a system with a precisely given energy and fixed composition (precise number of particles). The microcanonical ensemble contains with equal probability each possible state that is consistent with that energy and composition.
- Canonical ensemble describes a system of fixed composition that is in thermal equilibrium with a heat bath of a precise temperature. The canonical ensemble contains states of varying energy but identical composition; the different states in the ensemble are accorded different probabilities depending on their total energy.
- Grand canonical ensemble describes a system with non-fixed composition (uncertain particle numbers) that is in thermal and chemical equilibrium with a thermodynamic reservoir. The reservoir has a precise temperature, and precise chemical potentials for various types of particle. The grand canonical ensemble contains states of varying energy and varying numbers of particles; the different states in the ensemble are accorded different probabilities depending on their total energy and total particle numbers.

For systems containing many particles (the thermodynamic limit), all three of the ensembles listed above tend to give identical behaviour. It is then simply a matter of mathematical convenience which ensemble is used. The Gibbs theorem about equivalence of ensembles was developed into the theory of concentration of measure phenomenon, which has applications in many areas of science, from functional analysis to methods of artificial intelligence and big data technology.

Important cases where the thermodynamic ensembles *do not* give identical results include:

- Microscopic systems.
- Large systems at a phase transition.
- Large systems with long-range interactions.

In these cases the correct thermodynamic ensemble must be chosen as there are observable differences between these ensembles not just in the size of fluctuations, but also in average quantities such as the distribution of particles. The correct ensemble is that which corresponds to the way the system has been prepared and characterized—in other words, the ensemble that reflects the knowledge about that system.

	Thermodynamic ensembles		
Fixed variables	Microcanonical	Canonical	Grand canonical
	N, E, V	N, T, V	μ, T, V
Microscopic features	Number of microstates • W	Canonical partition function • $Z = \sum_k e^{-E_k/k_B T}$	Grand partition function • $\mathcal{Z} = \sum_k e^{-(E_k - \mu N_k)/k_B T}$
Macroscopic function	Boltzmann entropy • $S = k_B \log W$	Helmholtz free energy • $F = -k_B T \log Z$	Grand potential • $\Omega = -k_B T \log \mathcal{Z}$

Calculation Methods

Once the characteristic state function for an ensemble has been calculated for a given system, that system is 'solved' (macroscopic observables can be extracted from the characteristic state function). Calculating the characteristic state function of a thermodynamic ensemble is not necessarily a simple task, however, since it involves considering every possible state of the system. While some hypothetical systems have been exactly solved, the most general (and realistic) case is too complex for an exact solution. Various approaches exist to approximate the true ensemble and allow calculation of average quantities.

Exact

There are some cases which allow exact solutions.

- For very small microscopic systems, the ensembles can be directly computed by simply enumerating over all possible states of the system (using exact diagonalization in quantum mechanics, or integral over all phase space in classical mechanics).

- Some large systems consist of many separable microscopic systems, and each of the subsystems can be analysed independently. Notably, idealized gases of non-interacting particles have this property, allowing exact derivations of Maxwell–Boltzmann statistics, Fermi–Dirac statistics, and Bose–Einstein statistics.
- A few large systems with interaction have been solved. By the use of subtle mathematical techniques, exact solutions have been found for a few toy models. Some examples include the Bethe ansatz, square-lattice Ising model in zero field, hard hexagon model.

Monte Carlo

One approximate approach that is particularly well suited to computers is the Monte Carlo method, which examines just a few of the possible states of the system, with the states chosen randomly (with a fair weight). As long as these states form a representative sample of the whole set of states of the system, the approximate characteristic function is obtained. As more and more random samples are included, the errors are reduced to an arbitrarily low level.

- The Metropolis–Hastings algorithm is a classic Monte Carlo method which was initially used to sample the canonical ensemble.
- Path integral Monte Carlo, also used to sample the canonical ensemble.

Other

- For rarefied non-ideal gases, approaches such as the cluster expansion use perturbation theory to include the effect of weak interactions, leading to a virial expansion.
- For dense fluids, another approximate approach is based on reduced distribution functions, in particular the radial distribution function.
- Molecular dynamics computer simulations can be used to calculate microcanonical ensemble averages, in ergodic systems. With the inclusion of a connection to a stochastic heat bath, they can also model canonical and grand canonical conditions.
- Mixed methods involving non-equilibrium statistical mechanical results may be useful.

Non-equilibrium Statistical Mechanics

There are many physical phenomena of interest that involve quasi-thermodynamic processes out of equilibrium, for example:

- Heat transport by the internal motions in a material, driven by a temperature imbalance,
- Electric currents carried by the motion of charges in a conductor, driven by a voltage imbalance,
- Spontaneous chemical reactions driven by a decrease in free energy,
- Friction, dissipation, quantum decoherence,

- Systems being pumped by external forces (optical pumping, etc.),
- Irreversible processes in general.

All of these processes occur over time with characteristic rates, and these rates are of importance for engineering. The field of non-equilibrium statistical mechanics is concerned with understanding these non-equilibrium processes at the microscopic level. (Statistical thermodynamics can only be used to calculate the final result, after the external imbalances have been removed and the ensemble has settled back down to equilibrium.)

In principle, non-equilibrium statistical mechanics could be mathematically exact: ensembles for an isolated system evolve over time according to deterministic equations such as Liouville's equation or its quantum equivalent, the von Neumann equation. These equations are the result of applying the mechanical equations of motion independently to each state in the ensemble. Unfortunately, these ensemble evolution equations inherit much of the complexity of the underlying mechanical motion, and so exact solutions are very difficult to obtain. Moreover, the ensemble evolution equations are fully reversible and do not destroy information (the ensemble's Gibbs entropy is preserved). In order to make headway in modelling irreversible processes, it is necessary to consider additional factors besides probability and reversible mechanics.

Non-equilibrium mechanics is therefore an active area of theoretical research as the range of validity of these additional assumptions continues to be explored. A few approaches are described in the following subsections.

Stochastic Methods

One approach to non-equilibrium statistical mechanics is to incorporate stochastic (random) behaviour into the system. Stochastic behaviour destroys information contained in the ensemble. While this is technically inaccurate (aside from hypothetical situations involving black holes, a system cannot in itself cause loss of information), the randomness is added to reflect that information of interest becomes converted over time into subtle correlations within the system, or to correlations between the system and environment. These correlations appear as chaotic or pseudorandom influences on the variables of interest. By replacing these correlations with randomness proper, the calculations can be made much easier.

- *Boltzmann transport equation*: An early form of stochastic mechanics appeared even before the term "statistical mechanics" had been coined, in studies of kinetic theory. James Clerk Maxwell had demonstrated that molecular collisions would lead to apparently chaotic motion inside a gas. Ludwig Boltzmann subsequently showed that, by taking this molecular chaos for granted as a complete randomization, the motions of particles in a gas would follow a simple Boltzmann transport equation that would rapidly restore a gas to an equilibrium state.

  The Boltzmann transport equation and related approaches are important tools in non-equilibrium statistical mechanics due to their extreme simplicity. These approximations work well in systems where the "interesting" information is immediately (after just one collision) scrambled up into subtle correlations, which essentially restricts them to rarefied gases. The Boltzmann transport equation has been found to be very useful in simulations of electron transport in lightly doped semiconductors (in transistors), where the electrons are indeed analogous to a rarefied gas.

A quantum technique related in theme is the random phase approximation.

- *BBGKY hierarchy*: In liquids and dense gases, it is not valid to immediately discard the correlations between particles after one collision. The BBGKY hierarchy (Bogoliubov–Born–Green–Kirkwood–Yvon hierarchy) gives a method for deriving Boltzmann-type equations but also extending them beyond the dilute gas case, to include correlations after a few collisions.
- *Keldysh formalism* (a.k.a. NEGF—non-equilibrium Green functions): A quantum approach to including stochastic dynamics is found in the Keldysh formalism. This approach often used in electronic quantum transport calculations.

Near-equilibrium Methods

Another important class of non-equilibrium statistical mechanical models deals with systems that are only very slightly perturbed from equilibrium. With very small perturbations, the response can be analysed in linear response theory. A remarkable result, as formalized by the fluctuation-dissipation theorem, is that the response of a system when near equilibrium is precisely related to the fluctuations that occur when the system is in total equilibrium. Essentially, a system that is slightly away from equilibrium—whether put there by external forces or by fluctuations—relaxes towards equilibrium in the same way, since the system cannot tell the difference or "know" how it came to be away from equilibrium.

This provides an indirect avenue for obtaining numbers such as ohmic conductivity and thermal conductivity by extracting results from equilibrium statistical mechanics. Since equilibrium statistical mechanics is mathematically well defined and (in some cases) more amenable for calculations, the fluctuation-dissipation connection can be a convenient shortcut for calculations in near-equilibrium statistical mechanics.

A few of the theoretical tools used to make this connection include:

- Fluctuation–dissipation theorem
- Onsager reciprocal relations
- Green–Kubo relations
- Landauer–Büttiker formalism
- Mori–Zwanzig formalism.

Hybrid Methods

An advanced approach uses a combination of stochastic methods and linear response theory. As an example, one approach to compute quantum coherence effects (weak localization, conductance fluctuations) in the conductance of an electronic system is the use of the Green-Kubo relations, with the inclusion of stochastic dephasing by interactions between various electrons by use of the Keldysh method.

Macrostate

The macroscopic state of a thermodynamic system at equilibrium is specified by the values of a set of measurable thermodynamic parameters. For example, the macrostate of a fluid system can be specified by pressure, temperature and volume, (P, V, T). For an isolated system for which there is no exchange of energy or mass with the surroundings, the macrostate is specified by the internal energy E, number of particles N and volume V; (E, N, V). In case of a closed system which exchanges only energy with the surroundings (and no particle) and is in thermodynamic equilibrium with a heat bath at temperature T, the macrostate is given by (N, V, T). For an open system, exchange of both energy and particle with the surroundings can take place. For such systems, at equilibrium with a heat bath at temperature T and a pressure bath at pressure P (or a particle bath of chemical potential μ), the macrostate is specified by (N, P, T) (or (μ, V, T)).

The equilibrium of an isolated system corresponds to maximum entropy $S(E, N, V)$, for a closed system it corresponds to minimum Helmholtz free energy $F(N, V, T)$ and for an open system the equilibrium corresponds to minimum of Gibb's free energy $G(N, P, T)$ (or minimum of the grand potential $\Phi(\mu, V, T)$).

Microstate

A microstate of a system is obtained by specifying the states of all of its constituent elements. However, it depends on the nature of the constituent elements (or particles) of the system. Specification of microstates are made differently for classical and quantum particles.

Microstates of classical particles: In order to specify the microstates of a system of classical particles, one needs to specify the position q and the conjugate momentum p of each and every constituent particle of the system. In a classical system, the time evolution of q and p is governed by the classical Hamiltonian $H(p, q)$ and Hamilton's equation of motion,

$$\dot{q}_i = \frac{\partial H(p,q)}{\partial p_i} \text{ and } \dot{p}_i = -\frac{\partial H(p,q)}{\partial q_i}; \quad i = 1,2,\ldots,3N$$

For a system of N particles in 3-dimension. The state of a single particle at any time is then given by the pair of conjugate variables (q_i, p_i) a point in the phase space. Each single particle then constitutes a 6 -dimensional phase space (3 - coordinate and 3 - momentum). For N particles, the state of the system is then completely and uniquely defined by 3N canonical coordinates $q_1, q_2, \ldots, q_{3N}$, and 3N canonical momenta $p_1, p_2, \ldots, p_{3N}$. These 6N variables constitute a 6N - dimensional Γ - space or phase space of the system and each point of the phase space represents a microstate of the system. The locus of all the points in Γ - space satisfying the condition H(p, q) = E, total energy of the system, defines the energy surface.

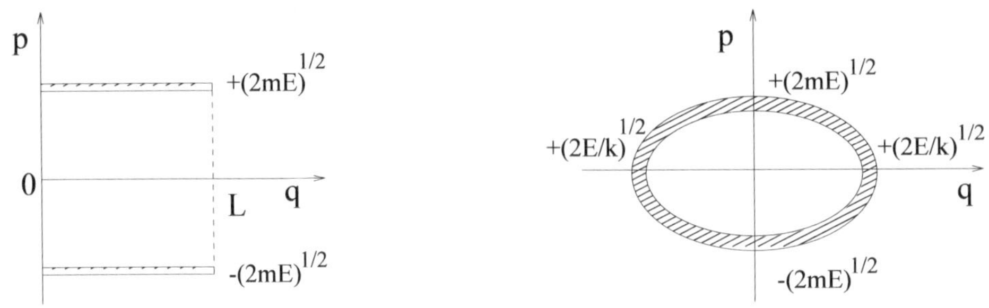

(a) (b)

(a) Accessible region of phase space for a free particle of mass m and energy E in a one dimensional box of length L. (b) Region of phase space for a one dimensional harmonic oscillator with energy E, mass m and spring constant k

Example: Consider a free particle of mass m inside a one dimensional box of length L, such that $0 \le q \le L$, with energy between E and E +δE. The macroscopic state of the system is defined by (E, N, L) with N =1. The microstates are specified in certain region of phase space. Since the energy of the particle is $E = p^2/2m$, the momentum will be $p = \pm\sqrt{2mE}$ and the position q is within 0 and L. However, there is a small width in energy δE, so the particles are confined in small strips of width δp = m/2EδE as shown in the figure above (a). Note that if δE = 0, the accessible region of phase space representing the system would be one dimensional in a two dimensional phase space. In order to avoid this artifact a small width in E is considered which does not affect the final results in the thermodynamic limit. In the figure above (b), the phase space region of a one dimensional harmonic oscillator with mass m, spring constant k and energy between E and E +δE is shown. The Hamiltonian of the particle is: $H = p^2/2m + kq^2/2$ and for a given energy E, the accessible region is an ellipse: $p^2/(2mE) + q^2(2E/k) = 1$. With the energy between E and E +δE, the accessible region is an elliptical shell of area $2\pi\sqrt{m/k}\delta E$.

Microstates of quantum particles: For a quantum particle, the state is characterized by the wave function $\psi(q_1, q_2, q_3, \ldots)$. Generally, the wave function is written in terms of a complete orthonormal basis of eigenfunctions of the Hamiltonian operator of the system. Thus, the wave function may be written as,

$$\psi \sum_n c_n \phi_n, \quad \hat{H}\phi_n = E_n \phi_n$$

where E_n is the eigenvalue corresponding to the state φ_n. The eigenstates φ_n, characterized by a set of quantum numbers n provides a way to count the microscopic states of the system.

Example: Consider a localized magnetic ion of spin 1/2 and magnetic moment μ in thermal equilibrium at temperature T. The particle has two eigenstates, (1, 0) and (0, 1) associated with spin up (↑) and down spin (↓) respectively. In the presence of an external magnetic field H ρ, the energy is given by,

$$E = -\vec{\mu} \cdot \vec{H} = \begin{cases} +\mu H, \text{ for spin} \downarrow \\ -\mu H, \text{ for spin} \uparrow \end{cases}$$

Thus, the system with macrostate (N, H, T) with N =1 has two microstates with energy $-\mu H$ and $+\mu H$

corresponding to up spin (parallel to H ρ) and down spin (antiparallel to $\vec{H}$). If there are two such magnetic ions in the system, it will have four microstates: ↑↑ with energy $-2\mu H$, ↑↓ & ↓↑ with zero energy and ↓↓ with energy $+2\mu H$. For a system of N spins of spin-1/2, there are total 2^N microstates and specification of the spin-states of all the N spins will give one possible microstate of the system.

Microscopic Definitions of Thermodynamic Concepts

Statistical mechanics links the empirical thermodynamic properties of a system to the statistical distribution of an ensemble of microstates. All macroscopic thermodynamic properties of a system may be calculated from the partition function that sums the energy of all its microstates.

At any moment a system is distributed across an ensemble of N microstates, each denoted by i, and having a probability of occupation p_i, and an energy E_i. If the microstates are quantum-mechanical in nature, then these microstates form a discrete set as defined by quantum statistical mechanics, and E_i is an energy level of the system.

Internal Energy

The internal energy of the macrostate is the mean over all microstates of the system's energy,

$$U = \langle E \rangle = \sum_{i=1}^{N} p_i E_i .$$

This is a microscopic statement of the notion of energy associated with the first law of thermodynamics.

Entropy

For the more general case of the canonical ensemble, the absolute entropy depends exclusively on the probabilities of the microstates and is defined as,

$$S = -k_B \sum_i p_i \ln p_i ,$$

where k_B is Boltzmann constant. For the microcanonical ensemble, consisting of only those microstates with energy equal to the energy of the macrostate, this simplifies to,

$$S = k_B \ln W ,$$

where W is the number of microstates. This form for entropy appears on Ludwig Boltzmann's gravestone in Vienna.

The second law of thermodynamics describes how the entropy of an isolated system changes in time. The third law of thermodynamics is consistent with this definition, since zero entropy means that the macrostate of the system reduces to a single microstate.

Heat and Work

Heat and work can be distinguished if we take the underlying quantum nature of the system into account.

For a closed system (no transfer of matter), heat in statistical mechanics is the energy transfer associated with a disordered, microscopic action on the system, associated with jumps in occupation numbers of the quantum energy levels of the system, without change in the values of the energy levels themselves.

Work is the energy transfer associated with an ordered, macroscopic action on the system. If this action acts very slowly, then the adiabatic theorem of quantum mechanics implies that this will not cause jumps between energy levels of the system. In this case, the internal energy of the system only changes due to a change of the system's energy levels.

The microscopic, quantum definitions of heat and work are the following:

$$\delta W = \sum_{i=1}^{N} p_i \, dE_i$$

$$\delta Q = \sum_{i=1}^{N} E_i \, dp_i$$

so that

$$dU = \delta W + \delta Q.$$

The two above definitions of heat and work are among the few expressions of statistical mechanics where the thermodynamic quantities defined in the quantum case find no analogous definition in the classical limit. The reason is that classical microstates are not defined in relation to a precise associated quantum microstate, which means that when work changes the total energy available for distribution among the classical microstates of the system, the energy levels (so to speak) of the microstates do not follow this change.

Correlation Function

Correlation Function in Statistical Mechanics

A function describing the influence of particles or groups of particles on one another and the effects due to the interaction of subsystems of the system under consideration.

In classical statistical mechanics, the correlation functions $G_2(1,2), G_3(1,2,3),...,$ are defined by the relations,

$$F_2(1,2), = F_1(1)F_1(2) + G_2(1,2),$$
$$F_3(1,2,3) = F_1(1)F_1(2)F_1(3) + F_1(1)G_2(2,3) +$$
$$+F_1(2)G_2(1,3) + F_1(3)F_2(1,2) + G_3(1,2,3),...,$$

where the symbols $1, 2,...,$ in the arguments of the functions denote the sets of coordinates r and momenta P of the 1st, 2nd ,..., particles, respectively, and $F_s(1,...,s)$ are the reduced distribution functions,

$$F_s(1,\ldots,s)=V\left(1-\frac{1}{N}\right)\ldots$$

$$\ldots\left(1-\frac{s-1}{N}\right)\int D_t d(s+1)\ldots dN,$$

where V is the volume of the system, N is the number of particles and the $D_t = D_t(1-,\ldots,N)$ are the distribution functions in the phase space at time t, normalized so that,

$$\int D_t(1,\ldots,N)\,d1\ldots dN = 1.$$

The variation of D_t in time is characterized by the Liouville equation $\partial D_t/\partial t = \wedge D_t$, where $\wedge$ represents the Liouville operator, which is not explicitly dependent on time. One usually considers the case in which $\wedge$ is the sum of an additive part and a binary part characterizing the interactions of the particles:

$$\wedge = \sum_{1\le j\le N}\wedge(j) + \sum_{1\le j_1\le j_2\le N}\wedge(j_1, j_2).$$

According to the principle of correlation damping, the correlation functions satisfy the boundary conditions $G_s(1,\ldots,s)\to 0$ as $\max\{|r_1-r_2|,\ldots,|r_1-r_s|,\ldots,|r_{s-1}r_s|\}\to\infty$.

The correlation functions $G_1(1)=F_1(1), G_2(1,2),\ldots,G_s(1,\ldots,s)$ are the functional derivatives,

$$G_s(1,\ldots,s)=\left(\frac{\delta^s A_t(u)}{\delta u(1)\delta u(2)\ldots\delta u(s)}\right)_{u=0},$$

of a functional $A_t(u)$ which is related to the so-called generating functional,

$$L_t(u)=\int\left\{\prod_{1\le j\le N}\left(1+\frac{V}{N}u(j)\right)\right\}D_t d1\ldots dN$$

by the relation,

$$L_t(u)=e^{A_t(u)}.$$

The functional $A_t(u)$ satisfies the equation,

$$\frac{\partial A_t(u)}{\partial t}=\int u(1)\wedge(1)\frac{\delta A_t(u)}{\delta u(1)}d1+$$

$$+\frac{1}{2}\int\left\{u(1)u(2)+\frac{N}{V}u(1)+\frac{N}{V}u(2)\right\}\wedge(1,2)$$

$$\left\{\frac{\delta A_t(u)}{\delta u(1)}\frac{\delta A_t(u)}{\delta u(2)}+\frac{\delta^2 A_t(u)}{\delta u(1)\delta u(2)}\right\}d1d2.$$

In quantum statistical mechanics, the correlation functions are operator quantities, defined as follows:

$$F_2(1,2) = S(1,2)\{F_1(1)F_1(2)\} + G_2(1,2),$$
$$F_3(1,2,3) = S(1,2,3)\{F_1(1)F_1(2)F_1(3)\} +$$
$$+\frac{1}{2}s(1,2,3)\{F_1(1)G_2(2,3) + F_1(2)G_2(1,3) +$$
$$+F_1(3)G_2(1,2)\} + G_3(1,2,3),\ldots,$$

where $S(1,2), S(1,2,3)$ are the symmetrization operator for Bose systems and the anti-symmetrization operator for Fermi systems. The correlation functions (*), forming the density matrix, satisfy the quantum-mechanical Liouville equation.

In quantum statistical mechanics, besides the correlation function (*) one considers correlation functions based on conventional thermodynamical averages, and correlation functions based on quasi-averages.

Bilinear combinations of correlation functions (both quantum-mechanical and classical) yield the Green functions. Correlation functions possess spectral representations; they satisfy the Bogolyubov inequality and a variation of the mean-value theorem.

Correlation functions corresponding to the Kirkwood decomposition are sometimes used; another version is a space-time correlation function.

Correlation functions may be interpreted as characteristic functions of probability measures.

Measuring Correlation Functions

Correlation functions are typically measured with scattering experiments. For example, x-ray scattering experiments directly measure electron-electron equal-time correlations.From knowledge of elemental structure factors, one can also measure elemental pair correlation functions. Equal-time spin–spin correlation functions are measured with neutron scattering as opposed to x-ray scattering. Neutron scattering can also yield information on pair correlations as well. For systems composed of particles larger than about one micrometer, optical microscopy can be used to measure both equal-time and equal-position correlation functions. Optical microscopy is thus common for colloidal suspensions, especially in two dimensions.

Time Evolution of Correlation Functions

In 1931, Lars Onsager proposed that the regression of microscopic thermal fluctuations at equilibrium follows the macroscopic law of relaxation of small non-equilibrium disturbances. This is known as the *Onsager regression hypothesis*. As the values of microscopic variables separated by large timescales, , should be uncorrelated beyond what we would expect from thermodynamic equilibrium, the evolution in time of a correlation function can be viewed from a physical standpoint as the system gradually 'forgetting' the initial conditions placed upon it via the specification of some microscopic variable. There is actually an intuitive connection between the time evolution of correlation functions and the time evolution of macroscopic systems: on average, the correlation function evolves in time in the same manner as if a system was prepared in the conditions specified by the correlation function's initial value and allowed to evolve.

Connection between Phase Transitions and Correlation Functions

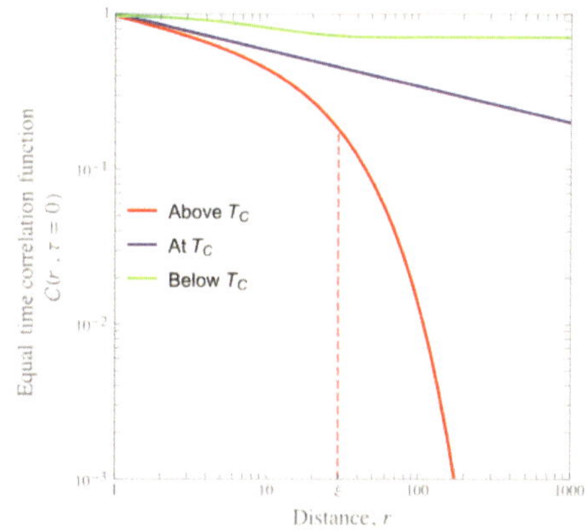

Equal-time correlation functions, $C(r,\tau=0)$, as a function of radius for a ferromagnetic spin system above, at, and below at its critical temperature, . Above T_C, $C(r,\tau=0)$ exhibits a combined exponential and power-law dependence on distance: $C(r,\tau=0) \propto r^{-(d-2+\eta)} e^{-r/\xi(T)}$. The power-law dependence dominates at distances short relative to the correlation length, ξ, while the exponential dependence dominates at distances large relative to ξ. At T_C, the correlation length diverges, $\xi(T_C)=\infty$, resulting in solely power-law behavior: $C(r,\tau=0) \propto r^{-(d-2+\eta)}$. T_C is distinguished by the extreme non-locality of the spatial correlations between microscopic values of the relevant order parameter without long-range order. Below T_C, the spins exhibit spontaneous ordering but finite correlation length again. Continuous order-disorder transitions can be understood as the process of the correlation length, ξ, transitioning from being finite in the low-temperature, ordered state, to infinite at the critical point, and then again finite in a high-temperature, disordered state.

Continuous phase transitions, such as order-disorder transitions in metallic alloys and ferromagnetic-paramagnetic transitions, involve a transition from an ordered to a disordered state. In terms of correlation functions, the equal-time correlation function is non-zero for all lattice points below the critical temperature, and is non-negligible for only a fairly small radius above the critical temperature. As the phase transition is continuous, the length over which the microscopic variables are correlated, ξ, must transition continuously from being infinite to finite when the material is heated through its critical temperature. This gives rise to a power-law dependence of the correlation function as a function of distance at the critical point. This is shown in the figure in the left for the case of a ferromagnetic material, with the quantitative details listed in the section on magnetism.

Applications

Magnetism

In a spin system, the equal-time correlation function is especially well-studied. It describes the canonical ensemble (thermal) average of the scalar product of the spins at two lattice points over all possible orderings: $G(r) = \langle s(R) \cdot s(R+r) \rangle - \langle s(R) \rangle \langle s(R+r) \rangle$. Here the brackets mean the above-mentioned thermal average. Schematic plots of this function are shown for a ferromagnetic material below, at, and above its Curie temperature on the left.

Even in a magnetically disordered phase, spins at different positions are correlated, i.e., if the distance r is very small (compared to some length scale ξ), the interaction between the spins will cause them to be correlated. The alignment that would naturally arise as a result of the interaction between spins is destroyed by thermal effects. At high temperatures exponentially-decaying correlations are observed with increasing distance, with the correlation function being given asymptotically by,

$$G(r) \approx \frac{1}{r^{d-2+\eta}} exp\left(\frac{-r}{\xi}\right),$$

where r is the distance between spins, and d is the dimension of the system, and η is a critical exponent. At high temperatures, the correlation decays to zero exponentially with the distance between the spins. The same exponential decay as a function of radial distance is also observed below T_C, but with the limit at large distances being the mean magnetization $\langle M^2 \rangle$.

As the temperature is lowered, thermal disordering is lowered, and in a continuous phase transition the correlation length diverges, as the correlation length must transition continuously from a finite value above the phase transition, to infinite below the phase transition:

$$\xi \propto |T - T_c|^{-\nu},$$

with another critical exponent.

This power law correlation is responsible for the scaling, seen in these transitions. All exponents mentioned are independent of temperature. They are in fact universal, i.e. found to be the same in a wide variety of systems.

Radial Distribution Functions

One common correlation function is the radial distribution function which is seen often in statistical mechanics. The correlation function can be calculated in exactly solvable models (one-dimensional Bose gas, spin chains, Hubbard model) by means of Quantum inverse scattering method and Bethe ansatz. In an isotropic XY model, time and temperature correlations were evaluated by Its, Korepin, Izergin & Slavnov.

Higher Order Correlation Functions

Higher-order correlation functions involve multiple reference points, and are defined through a generalization of the above correlation function by taking the expected value of the product of more than two random variables:

$$C_{i_1 i_2 \cdots i_n}(s_1, s_2, \cdots, s_n) = \langle X_{i_1}(s_1) X_{i_2}(s_2) \cdots X_{i_n}(s_n) \rangle.$$

However, such higher order correlation functions are relatively difficult to interpret and measure. For example, in order to measure the higher-order analogues of pair distribution functions, coherent x-ray sources are needed. Both the theory of such analysis and the experimental measurement of the needed X-ray cross-correlation functions are areas of active research.

Partition Function

The partition function is an important quantity in statistical mechanics which encodes the statistical properties of a system in thermodynamic equilibrium. This equation is particular to the canonical ensemble, that is the ensemble in which the system of interest is allowed to exchange heat with its surrounding environment, with a fixed temperature, volume and number of particles. There is also the grand canonical ensemble, in which the system can exchange heat and particles with the environment, with a fixed temperature, volume and chemical potential. The partition function for this ensemble is,

$$\Xi = \sum_{r} \sum_{N} e^{\frac{-(r_{\Gamma} - \mu N)}{kT}}$$

Canonical Partition Function

As a beginning assumption, assume that a thermodynamically large system is in constant thermal contact with the environment, with a temperature *T*, and both the volume of the system and the number of constituent particles fixed. This kind of system is called a canonical ensemble. Let us label with s ($s = 1, 2, 3, \ldots$) the *exact* states (microstates) that the system can occupy, and denote the total energy of the system when it is in microstate s as E_s. Generally, these microstates can be regarded as analogous to discrete quantum states of the system.

The canonical partition function is,

$$Z = \sum_{s} e^{-\beta E_s},$$

where the "inverse temperature", β, is conventionally defined as,

$$\beta \equiv \frac{1}{k_B T}$$

with k_B denoting Boltzmann's constant. The term $\exp(-\beta \cdot E_s)$ is known as the Boltzmann factor. In systems with multiple quantum states s sharing the same E_s, it is said that the energy levels of the system are degenerate. In the case of degenerate energy levels, we can write the partition function in terms of the contribution from energy levels (indexed by j) as follows:

$$Z = \sum_{j} g_j \cdot e^{-\beta E_j},$$

where g_j is the degeneracy factor, or number of quantum states s which have the same energy level defined by $E_j = E_s$.

The above treatment applies to *quantum* statistical mechanics, where a physical system inside a finite-sized box will typically have a discrete set of energy eigenstates, which we can use as the states s above. In *classical* statistical mechanics, it is not really correct to express the partition function as a sum of discrete terms, as we have done. In classical mechanics, the position and momentum variables of a particle can vary continuously, so the set of microstates is actually uncountable. In this case we must describe the partition function using an integral rather than a sum. For instance, the partition function of a gas of N identical classical particles is,

$$Z = \frac{1}{N!h^{3N}} \int exp[-\beta H(p_1 \cdots p_N, x_1 \cdots x_N)] \, d^3p_1 \cdots d^3p_N \, d^3x_1 \cdots d^3x_N$$

where,

p_i indicate particle momenta

x_i indicate particle positions

d^3 is a shorthand notation serving as a reminder that the p_i and x_i are vectors in three dimensional space, and

H is the classical Hamiltonian.

The results will apply equally well to the continuous form. The extra constant factor introduced in the denominator was introduced because, unlike the discrete form, the continuous form shown above is not dimensionless. To make it into a dimensionless quantity, we must divide it by h^{3N} where h is some quantity with units of action (usually taken to be Planck's constant).

In quantum mechanics, the partition function can be more formally written as a trace over the state space (which is independent of the choice of basis):

$$Z = tr(e^{-\beta \hat{H}}),$$

where $\hat{H}$ is the quantum Hamiltonian operator. The exponential of an operator can be defined using the exponential power series. The classical form of Z is recovered when the trace is expressed in terms of coherent states and when quantum-mechanical uncertainties in the position and momentum of a particle are regarded as negligible. Formally, one inserts under the trace for each degree of freedom a resolution of the identity,

$$\mathbf{1} = \int |x, p\rangle\langle x, p| \frac{dx dp}{h}$$

where $|x, p\rangle$ is a normalised Gaussian wavepacket centered at position x and momentum p. Thus,

$$Z = \int tr\left(e^{-\beta \hat{H}} |x, p\rangle\langle x, p|\right) \frac{dx dp}{h} = \int \langle x, p| e^{-\beta \hat{H}} |x, p\rangle \frac{dx dp}{h}$$

A coherent state is an approximate eigenstate of both operators $\hat{x}$ and $\hat{p}$, hence also of the Hamiltonian $\hat{H}$, with errors of the size of the uncertainties. If Δx and Δp can be regarded as zero, the action of $\hat{H}$ reduces to multiplication by the classical Hamiltonian, and Z reduces to the classical configuration integral.

Significance

It may not be obvious why the partition function, as we have defined it above, is an important quantity. First, let us consider what goes into it. The partition function is a function of the temperature T and the microstate energies E_1, E_2, E_3, etc. The microstate energies are determined by other thermodynamic variables, such as the number of particles and the volume, as well as microscopic quantities like the mass of the constituent particles. This dependence on microscopic variables is the central point of statistical mechanics. With a model of the microscopic constituents of a system, one can calculate the microstate energies, and thus the partition function, which will then allow us to calculate all the other thermodynamic properties of the system.

The partition function can be related to thermodynamic properties because it has a very important statistical meaning. The probability P_s that the system occupies microstate s is,

$$P_s = \frac{1}{Z} e^{-\beta E_s}.$$

$e^{-\beta E_s}$ is the well-known Boltzmann factor. The partition function thus plays the role of a normalizing constant (note that it does *not* depend on s), ensuring that the probabilities sum up to one:

$$\sum_s P_s = \frac{1}{Z} \sum_s e^{-\beta E_s} = \frac{1}{Z} Z = 1.$$

This is the reason for calling Z the "partition function": it encodes how the probabilities are partitioned among the different microstates, based on their individual energies. The letter Z stands for the German word *Zustandssumme*, "sum over states". This notation also implies another important meaning of the partition function of a system: it counts the (weighted) number of states a system can occupy. Hence if all states are equally probable (equal energies) the partition function is the total number of possible states. Often this is the practical importance of Z.

Calculating the Thermodynamic Total Energy

In order to demonstrate the usefulness of the partition function, let us calculate the thermodynamic value of the total energy. This is simply the expected value, or ensemble averagefor the energy, which is the sum of the microstate energies weighted by their probabilities:

$$\langle E \rangle = \sum_s E_s P_s = \frac{1}{Z} \sum_s E_s e^{-\beta E_s} = -\frac{1}{Z} \frac{\partial}{\partial \beta} Z(\beta, E_1, E_2, \cdots) = -\frac{\partial \ln Z}{\partial \beta}$$

or, equivalently,

$$\langle E \rangle = k_B T^2 \frac{\partial \ln Z}{\partial T}.$$

Incidentally, one should note that if the microstate energies depend on a parameter λ in the manner

$$E_s = E_s^{(o)} + \lambda A_s \qquad \text{for all } s$$

then the expected value of A is,

$$\langle A\rangle=\sum_s A_s P_s=-\frac{1}{\beta}\frac{\partial}{\partial\lambda}\ln Z(\beta,\lambda).$$

This provides us with a method for calculating the expected values of many microscopic quantities. We add the quantity artificially to the microstate energies (or, in the language of quantum mechanics, to the Hamiltonian), calculate the new partition function and expected value, and then set λ to zero in the final expression. This is analogous to the source fieldmethod used in the path integral formulation of quantum field theory.

Relation to Thermodynamic Variables

In this section, we will state the relationships between the partition function and the various thermodynamic parameters of the system. These results can be derived using the method of the previous section and the various thermodynamic relations.

As we have already seen, the thermodynamic energy is,

$$\langle E\rangle=-\frac{\partial\ln Z}{\partial\beta}.$$

The variance in the energy (or "energy fluctuation") is,

$$\langle(\Delta E)^2\rangle\equiv\langle(E-\langle E\rangle)^2\rangle=\frac{\partial^2\ln Z}{\partial\beta^2}.$$

The heat capacity is,

$$C_v=\frac{\partial\langle E\rangle}{\partial T}=\frac{1}{k_BT^2}\langle(\Delta E)^2\rangle.$$

The entropy is,

$$S\equiv-k_B\sum_s P_s\ln P_s=k_B(\ln Z+\beta\langle E\rangle)=\frac{\partial}{\partial T}(k_BT\ln Z)=-\frac{\partial A}{\partial T}$$

where A is the Helmholtz free energy defined as $A = U - TS$, where $U=<E>$ is the total energy and S is the entropy, so that,

$$A=\langle E\rangle-TS=-k_BT\ln Z.$$

Partition Functions of Subsystems

Suppose a system is subdivided into N sub-systems with negligible interaction energy. If the partition functions of the sub-systems are $\zeta_1, \zeta_2, \ldots, \zeta_N$, then the partition function of the entire system is the *product* of the individual partition functions:

$$Z=\prod_{j=1}^{N}\zeta_j$$

If the sub-systems have the same physical properties, then their partition functions are equal, $\zeta_1 = \zeta_2 = \ldots = \zeta$, in which case,

$$Z = \zeta^N.$$

However, there is a well-known exception to this rule. If the sub-systems are actually identical particles, in the quantum mechanical sense that they are impossible to distinguish even in principle, the total partition function must be divided by a $N!$ (N factorial):

$$Z = \frac{\zeta^N}{N!}.$$

This is to ensure that we do not "over-count" the number of microstates. While this may seem like a strange requirement, it is actually necessary to preserve the existence of a thermodynamic limit for such systems. This is known as the Gibbs paradox.

Grand Canonical Partition Function

In a manner similar to the definition of the canonical partition function for the canonical ensemble, we can define a grand canonical partition function for a grand canonical ensemble, a system that can exchange both heat and particles with the environment, which has a constant temperature T, and a chemical potential μ. The grand canonical partition function, although conceptually more involved, simplifies the theoretical handling of quantum systems because it incorporates in a simple way the spin-statistics of the particles (i.e. whether particles are bosons or fermions). A canonical partition function, which has a given number of particles is in fact difficult to write down because of spin statistics.

The grand canonical partition function $\mathcal{Z}$ for an ideal quantum gas (a gas of non-interacting particles in a given potential well) is given by the following expression:

$$\mathcal{Z} = \sum_{N=0}^{\infty} \sum_{\{n_i\}} \prod_i e^{-\beta n_i(\epsilon_i - \mu)}$$

where N is the total number of particles in the gas, index i runs over every microstate (that is, a single particle state in the potential) with n_i being the number of particles occupying microstate i and ε_i being the energy of a particle in that microstate. The set $\{ n_i \}$ is the collection of all possible occupation numbers for each of these microstates such that $\Sigma\, n_i = N$.

For example, consider the $N = 3$ term in the above sum. One possible set of occupation numbers would be $n_i = 0, 1, 0, 2, 0 \ldots$ and the contribution of this set of occupation numbers to the $N = 3$ term would be,

$$\prod_i e^{-\beta n_i(\epsilon_i - \mu)} = e^{-\beta(\epsilon_1 - \mu)} e^{-2\beta(\epsilon_3 - \mu)}.$$

For bosons, the occupation numbers can take any integer values as long as their sum is equal to N. For fermions, the Pauli exclusion principle requires that the occupation numbers only be 0 or 1, again adding up to N.

Probability

Specific Expressions

The expression for the grand partition function can be shown to be mathematically equivalent to:

$$\mathcal{Z} = \prod_i \mathcal{Z}_i \,.$$

(The above product is sometimes taken over all states with equal energy, rather than over each state, in which case the individual partition functions must be raised to a power g_i where g_i is the number of such states. g_i is also referred to as the "degeneracy" of states.)

For a system composed of bosons:

$$= \sum^{\infty} \quad {}^{-\beta n_i(\epsilon_i-\mu)} = \frac{}{{}^{-\beta(\epsilon\ -\mu)}}$$

and for a system composed of fermions:

$$= \sum e^{-\beta n_i(\epsilon_i-\mu)} = \quad + e^{-\beta(\epsilon_i-\mu)} \,.$$

For the case of a Maxwell-Boltzmann gas, we must use "correct Boltzmann counting" and divide the Boltzmann factor $e^{-\beta(\epsilon_i-\mu)}$ by n_i.

$$\mathcal{Z}_i = \sum_{n_i=0}^{\infty} \frac{e^{-\beta n_i(\epsilon_i-\mu)}}{n_i\,!} = exp\big(exp(-\beta(\epsilon_i-\mu))\big) \,.$$

Relation to Thermodynamic Variables

Just as with the canonical partition function, the grand canonical partition function can be used to calculate thermodynamic and statistical variables of the system. As with the canonical ensemble, the thermodynamic quantities are not fixed, but have a statistical distribution about a mean or expected value.

Occupation Numbers

The most probable occupation numbers are:

$$\langle n_i \rangle = -\left(\frac{\partial ln(\mathcal{Z}_i)}{\partial \alpha}\right)_{\beta,V} = \frac{1}{\beta}\left(\frac{\partial ln(\mathcal{Z}_i)}{\partial \mu}\right)_{\beta,V} ,$$

where $\alpha = -\beta \cdot \mu$.

For Boltzmann particles this yields:

$$\langle n_i \rangle = e^{-\beta(\epsilon_i-\mu)} \,.$$

For bosons:

$$\langle n_i \rangle = \frac{1}{e^{\beta(\epsilon_i-\mu)} - 1} \,.$$

For fermions:

$$\langle n_i \rangle = \frac{1}{e^{\beta(\epsilon_i - \mu)} + 1} .$$

which are just the results found using the canonical ensemble for Maxwell-Boltzmann statistics, Bose-Einstein statistics and Fermi-Dirac statistics, respectively. (The degeneracy g_i is missing from the above equations because the index i is summing over individual microstates rather than energy eigenvalues.)

Total number of particles:

$$\langle N \rangle = -\left(\frac{\partial ln(\mathcal{Z})}{\partial \alpha}\right)_{\beta,V} = \frac{1}{\beta}\left(\frac{\partial ln(\mathcal{Q})}{\partial \mu}\right)_{\beta,V} .$$

Variance in total number of particles:

$$\langle (\delta N)^2 \rangle = \left(\frac{\partial^2 ln(\mathcal{Z})}{\partial \alpha^2}\right)_{\beta,V} .$$

Internal energy:

$$\langle E \rangle = -\left(\frac{\partial ln(\mathcal{Z})}{\partial \beta}\right)_{\mu,V} + \mu\langle N \rangle .$$

Variance in internal energy:

$$\langle (\delta E)^2 \rangle = \left(\frac{\partial^2 ln(\mathcal{Z})}{\partial \beta^2}\right)_{\mu,V} .$$

Pressure:

$$P = k_B T\left(\frac{\partial ln(\mathcal{Z})}{\partial V}\right)_N .$$

Mechanical equation of state:

$$\langle PV \rangle = \frac{ln\,\mathcal{Z}}{\beta} .$$

Relation to Potential V

For the case of a non-interacting gas, using the "Semiclassical Approach" we can write (approximately) the inverse of the potential in the form:

$$\frac{1}{2i\pi}\int_{c-i\infty}^{c+i\infty} \mathrm{ds}\frac{Z(s)}{\sqrt{\pi s}}e^{st} \sim V^{-1}(t)$$

$$s = \frac{1}{k_B T} \text{ (valid for high T)}$$

supposing that the Hamiltonian of every particle is H=T+V.

Finite Quantum Grand Canonical Ensemble

If the number of particles N is maintained small and finite, the orthodicity of the canonical ensemble is exploited to define a finite quantum grand canonical partition function for a finite quantum grand canonical ensemble. The only hypothesis in addition is that the system is ergodic, so the average in time is equivalent to the average on the possible configurations. A system is orthodic if the averages of the physical quantities on a given distribution recover the laws of thermodynamics. The canonical ensemble is natively orthodic so such properties is granted for the finite N grand canonical ensemble by construction. Said $H(N)$ the Hamiltonian of the system constituted by N electrons distributed between a two dimensional quantum dot of area Σ_1 containing only one or two electrons and a two dimensional finite and small gas of area Σ_2, it can be separated by $H = H_1(N_1) + H_2(N_2)$ where $N_1 = 1,2$. Consequently one defines the finite quantum grand partition function for identical particles from:

$$\mathcal{Z}(\Sigma, N, T) = \sum_{N_1=1}^{2} tr(e^{-\beta \widehat{H_1}})$$

$$\mathcal{Z} = e^{-c_A \Sigma_1 / k_B}$$

where the heat capacity per area unit at constant surface c_A is:

$$c_A = \frac{1}{A}\left(\frac{\partial U(A,N)}{\partial T}\right)_A = \frac{\pi^2}{3} g k_B^2 T$$

where the density of states per surface unit is $g(E)dE = m\pi^2 / \hbar^2 dE$ for a two dimensional system (d = 2).

The generalized temperature is obtained by considering time ensembles instead of configuration space ensembles in the case of thermal and particle exchange between a small system of fermions (N even less than 10) with a single/double occupancy system which acts as single particle thermometer. The generalized temperature is expressed as the ratio of the average time of occupation τ_1 and τ_2 of the single/double occupancy system:

$$T = k^{-1} \ln 2 \frac{\tau_2}{\tau_1}\left(E - E_F\left(1 + \frac{3}{2N}\right)\right),$$

where E_F is the Fermi energy which tends to the ordinary temperature when N goes to infinity.

Statistical Ensemble

A name given in statistical physics (cf. Statistical physics, mathematical problems in) to the phase space (space of states) of any physical system and a method of averaging physical (observable) quantities related to this system. In a classical system with phase space Ω, the observable quantities are real functions defined on Ω, and they are averaged by integration with respect to a certain probability measure μ on Ω. In a quantum system described by vectors in a Hilbert space $\mathcal{H}$, the observable quantities are defined by self-adjoint operators acting on $\mathcal{H}$, and are averaged using a certain positive, normalized functional ρ, defined on the algebra $\mathfrak{A}(\mathcal{H})$ of operators on $\mathcal{H}$ (such functionals on $\mathfrak{A}(\mathcal{H}\Rightarrow$ are called states). A state is usually given in the form,

$$\rho(A)=\mathrm{Tr}(\widehat{\rho}A), \qquad A\in\mathfrak{A}(\mathcal{H}),$$

where $\hat{p}$ is a positive operator on $\mathcal{H}$ (of trace class) such that Tr $\hat{p}$ (the operator $\hat{p}$ is called the density matrix of the state p).

If the evolution of the physical system in time is given (the dynamics of the system), i.e. (in the classical case) a group If the evolution of the physical system in time is given (the dynamics of the system), i.e. (in the classical case) a group $\Gamma_t:\Omega\to\Omega,\ t\in\mathrm{R}$, of one-to-one mappings of the phase space onto itself, generated by the Hamilton equations of motion with a certain Hamilton function $H(\omega),\omega\in\Omega$ (the energy of the system), or (in the quantum case) a group of unitary mappings $U_t:H\to H, t\in\mathrm{R}$, of the Hilbert space onto itself, generated by the Hamilton operator $\hat{H}$ (the energy operator of the system), then the evolution in time of any statistical ensemble defined for this system is naturally defined:

$$\left.\begin{aligned}&\mu_t(C)=\mu(\Gamma_t^{-1}C),\quad C\subset\Omega\ \text{(classical case)},\\ &\rho t(A)=\rho(U_t\,AU_t^{-1})\ \ \text{(quantum case)}.\end{aligned}\right\}$$

In describing the stationary behaviour of the system, equilibrium ensembles are studied, i.e. measures or states that are invariant relative to the evolution above. Although there are, generally speaking, many equilibrium statistical ensembles, only special ones are studied in statistical physics — the so-called canonical Gibbs ensembles (distributions).

Classical Gibbs Ensembles

In addition to the Hamiltonian $H=H_0(\omega)$, let there be a functionally independent set $H_1(w),...,H_K(\omega), k=0,1,...$, of functions on Ω that are invariant relative to the dynamics $\tilde{\mathrm{A}}_t$ (in systems consisting of a finite, but arbitrary, number of particles of one or more types, $H_i(\omega)$ is equal, for example, to the number of particles of whatever type in the configuration $\omega\in\Omega$; in the case of a system of magnetic dipoles, $H_i(\omega)$ is equal to their total magnetic moment, etc.). The measure,

$$d\,\mu\beta,v_1,...,v_k=$$

$$=(\Xi)^{-1}\exp\{-\beta(\mathrm{H}_0+{}_{v1}H1+...+{}_{vk}H_k)\}d\,w,$$

is called a grand canonical Gibbs ensemble, where $d\,\omega$ is a measure generated by a symplectic structure on Ω; $\beta\rangle 0$ and $v_1,...,v_k$ are real parameters; and Ξ is a normalization factor, called the grand statistical sum,

$$\Xi=\int_\Omega \exp\{-\beta(H_0+{}_{v1}H_1+...+{}_{vk}H_k)\}dw.$$

The measure $\mu h_1,...,h_k$ generated by the measure equation mentioned before the above one, concentrated on the set $\{\omega : H_i(\omega) = h_i,\ \square i = 0,...,k\}, h_i \in \mathrm{R}, i = 0,...,k$, and coinciding with the conditional distribution on this set is called a micro-canonical Gibbs ensemble. "Intermediate" ensembles are also studied — the so-called small canonical Gibbs ensembles, which are obtained in similar fashion from equation mentioned before the above one, by fixing the values of all or some of the functions $H_i(\omega), i = 1,..,k$.

Quantum Gibbs Ensembles

Let $\hat{H}_1,...,\hat{H}_k$ (k arbitrary) be pairwise commuting operators, commuting also with the operator $\hat{H} = \hat{H}_0$ ("conserved quantities"). The state on $\mathfrak{A}(\mathcal{H})$ defined by the density matrix,

$$\widehat{\rho}\,\beta, {}_{v_1,...,v_k} = (\Xi)^{-1} \exp\{-\beta(\widehat{H}_0 +_{v1} \widehat{H}_1 + ... +_{vk} \widehat{H}_k)\},$$

Where,

$$\Xi = \mathrm{Tr}(\exp\{-\beta(\widehat{H}_0 +_{v1} \widehat{H}_1 + ... +_{vk} \widehat{H}_k)\})$$

is a grand statistical sum (grand partition function) and $\beta \rangle 0, v_1,...,v_k$ are parameters, is called the grand canonical Gibbs ensemble. Let $P^{h_i}_{\hat{H}_i}, i = 0,...,k$, be a projector on $\mathcal{H}$ onto the eigensubspace of the operator $\hat{H}_i$ with eigenvalue h_i. A state with density matrix,

$$\hat{\rho}\,h_0,...,h_k = (Q)^{-1} P^{h_0}_{\hat{H}_0} ... P^{h_k}_{\hat{H}_k},$$

where $Q = \dim(\mathrm{Im}\hat{\rho}h_0,...,h_k)$, is called a micro-canonical Gibbs ensemble. Small canonical Gibbs ensembles are introduced in the same way, for example, as a statistical ensemble with density matrix,

$$\hat{\rho}\,\beta, h_1,...,h_k = (Z)^{-1} \exp\{-\beta\hat{H}_0\} \prod_{i=1}^{k} P^{h_i}_{\hat{H}_i},$$

where Z is a normalization factor (a small statistical sum, small partition function).

Certain model modifications of these statistical ensembles are sometimes studied (for example, Gibbs ensembles in configuration or lattice systems), as are so-called limiting Gibbs ensembles, i.e. probability distributions (or states) on the phase space of an infinite system (for example, systems with an infinite number of particles moving throughout the space). These statistical ensembles arise from the above ensembles for finite systems during a passage to the thermodynamic limit.

There exists a hypothesis — the so-called principle of limit equivalence of ensembles — which states that when certain natural conditions are fulfilled (roughly speaking, when there is no phase transition), limiting Gibbs statistical ensembles obtained using different Gibbs ensembles (grand canonical, small canonical, micro-canonical) coincide, given a specific correspondence between the parameters defining these ensembles. This hypothesis has been proved in certain specific situations.

Particle Statistics

In statistical mechanics, particle statistics is the description of a system of multiple particles. A fundamental concept within particle statistics is a statistical ensemble that describes the statistical properties of the system as a whole. Particle statistics are broadly differentiated into two classes, namely classical statistics and quantum statistics. In classical statistics, the particles in a system are considered distinguishable and governed by the Maxwell-Boltzmann statistics. In quantum statistics, the particles are indistinguishable and governed either by Bose-Einstein statistics or Fermi-Dirac statistics, depending on the particle studied.

References

- Mahon, Basil (2003). The Man Who Changed Everything – the Life of James Clerk Maxwell. Hoboken, NJ: Wiley. ISBN 978-0-470-86171-4. OCLC 52358254
- Statistical-mechanics, the-fundamentals-of-statistical-mechanics: physics.tutorvista.com, Retrieved 27 March 2018
- Touchette, H. Equivalence and nonequivalence of ensembles: Thermodynamic, macrostate, and measure levels. Journal of Statistical Physics, 159(5), 987-1016, 2015. doi:10.1007/s10955-015-1212-2
- Correlation-function-in-statistical-mechanics: encyclopediaofmath.org, Retrieved 28 May 2018
- Ledoux M. 2001 The Concentration of Measure Phenomenon. (Mathematical Surveys & Monographs No. 89). Providence: AMS. doi:10.1090/surv/089
- Statistical-ensemble: encyclopediaofmath.org, Retrieved 18 July 2018

Nuclear Physics: An Introduction

Nuclear physics studies the nucleus, its structure and interactions. The concepts of nuclear structure, stability of the nucleus, nuclear spin, radioactivity, etc. are essential for the development of the subject. Various other topics such as nuclear magnetic resonance, nuclear reactors, nuclear reactions, etc. have also been extensively covered for an extensive understanding of the subject.

The branch of science that deals with the fundamental particle namely nucleus is called Nuclear Physics. Nuclear physics basically deals with the structure and components of nucleus. The properties like radius, mass etc are dealt with and studied. Various models related to the nucleus and their development by various scientists, their experiments, theories etc. are the part of Nuclear Physics.

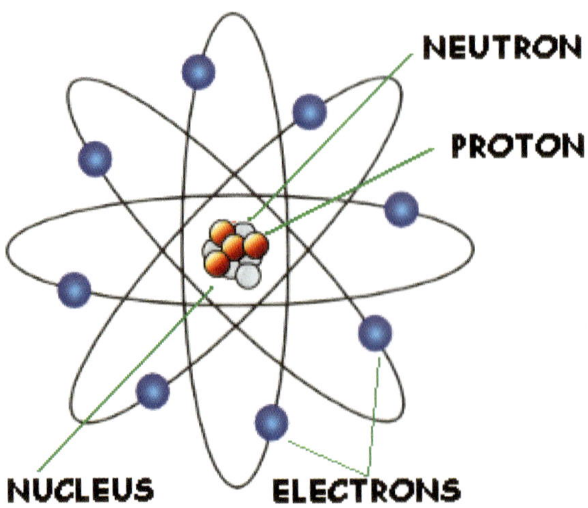

Nuclear physics basically deals with the structure and components of nucleus. The properties like radius, mass etc are dealt with and studied. Various models related to the nucleus and their development by various scientists, their experiments, theories etc. are the part of nuclear physics.

The interaction and constituents of nuclei are studied in nuclear physics. There are various applications of nuclear physics such as:

Nuclear power generation, Nuclear weapons technology, Nuclear medicine, Magnetic resonance imaging, Ion implantation in materials engineering, radiocarbon dating in geology and archaeology.

Ernest Rutherford is a very famous chemists. He is also known as Father of Nuclear physics as he was the first to propose the Nuclear model with a charged nucleus in the center. He found the nucleus by his very famous gold scattering experiment.

According to this Model,

1. All the mass of atom is concentrated within a small positively charged region at the center called the NUCLEUS.
2. The nucleus is surrounded by the electrons which are negatively charged that move around it in circular paths. These circular paths are called the ORBITS.
3. The Rutherford model resembles our solar system. Here electrons and the nucleus are held together by Electrostatic forces of attraction.

The Radius of nucleus denoted by R is given by,

$R = R_o A^{\frac{1}{3}}$

Here,

A is the mass number of the element,

(Mass number = Total no of protons + Total no of neutrons)

R_o = Proportionality Constant,

R is directly proportional to $A^{\frac{1}{3}}$,

Volume of nucleus = $\frac{4}{3}\pi R^3$

Thus V α A

His model developed a new idea for the scientists which made Bohr to give his Bohr's model for an atom. Thus he was awarded the Nobel Prize in Chemistry in 1908.

Nuclear Physics Equations

The Mass number (A), also called atomic mass number or nucleon number, is the total number of protons and neutrons in a nucleus.

$A = Z + N$

where A = Mass number,

Z = Proton number,

N = Neutron number.

When Proton and Neutron combines together in order to form nucleus then in this process some mass gets lost and this lost mass is known as Mass defect. More is the mass defect more is presumed to be stability of product nuclei.

Mathematically,

Δm = Mass of Nucleons – Mass of Nucleus.

$= A - M$

$= (Z\,m_p + N\,m_n) - M.$

$\Delta\,m = Z\,m_p + (A - Z)\,m_n - M$

Here M = Mass of the nucleus

Δ m = Mass of the nucleons - Mass of the nucleus.

m_p = Mass of the Proton and

m_n = Mass of the Neutron.

Packing fraction is defined as Mass defect per nucleon.

Packing fraction (f) = Mass defect per nucleons

$$= \frac{\Delta m}{A}$$

$$= \frac{M - A}{A}$$

$$\text{Packing Fraction (f)} = \frac{\left[Zm_p + (A - Z)m_n - A\right]}{A}$$

Greater the negative value of packing fraction higher is the stability of product nuclei.

Nuclear Binding Energy

In nucleus, proton and neutron are contained in small sphere. Due to such small small distance between two protons, they exert a very large force of repulsion on each other.

Therefore to prevent nucleus from bursting, certain amount of energy needed to bind the nucleus. This energy needed is known as Binding Energy.When nucleus is formed, the mass of nucleons forming it decreases. This mass defect provides the required binding energy. The energy equivalent to mass defect is called Binding Energy.

Binding energy of a nucleus may also be defined as the amount of work required to separate the nucleons at infinite distance. It is the measure of stability of nucleus.

Greater the binding energy per nucleons greater will be nucleus stability.

Rutherford Gold Foil Scattering

In 1910, a physicist from New Zealand, Ernest Rutherford performed an experiment known as Rutherford's Gold Foil Experiment. This experiment was determined to find out the structure of an atom.

Rutherford gave this atomic theory on the basis of following observations during the gold foil experiment:

1. Since most of the alpha particles passed through the foil without any deflection, thus it was concluded that most of the atom is empty.

2. As some particles deflected by small angles, it indicated that positively charged alpha particles were attracted by negatively charged particles means electrons.

3. Large angles of deflection experienced by the alpha particles indicated that there is a massive positively charged region present in the atom and due to repulsion alpha particles were deflected at large angles.

Compton Scattering

Compton effect was first observed by Arthur Compton in 1923. It occurs when the incident x-ray photon interacts with the electron which makes the incident photon to deflect away from its original path.

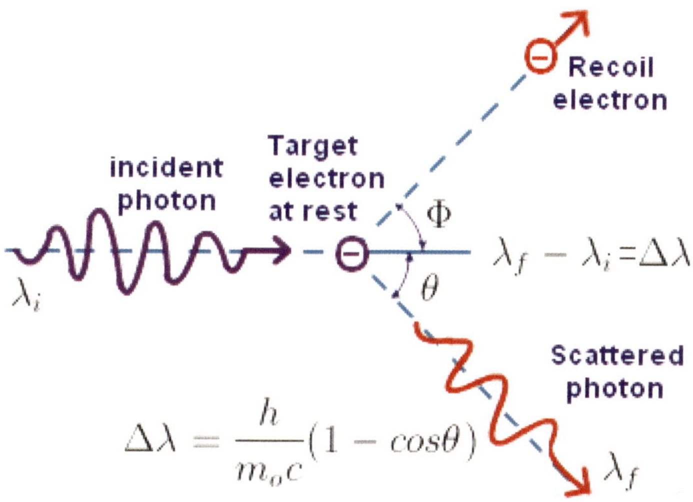

The change in wavelength is given by:

$$\lambda' - \lambda = \frac{h}{m_e c}(1 - \cos\theta)$$

Compton scattering is an inelastic scattering because the wavelength of the scattered light is different from that of the incident radiation. The change in the wavelength is called the Compton shift.

Photoelectric Effect

When light of suitable wavelength is incident on the metal plate electrons are emitted. These emitted electrons are called as photo electrons and the phenomenon is Photoelectric Effect.

Pair Production

When a photon is incident on the atomic nucleus, electron-positron pair is produced. This is called Pair-production. Pair production is a phenomenon of nature where energy is converted in to mass.

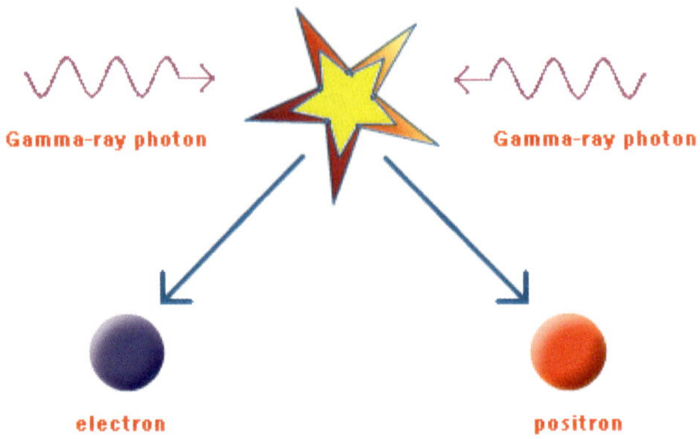

When two high energy gamma-ray photons collide, an electron/positron pair produced (the energy is converted into mass, $E=mc^2$)

Cherenkov Radiation

Cherenkov radiation is electromagnetic radiation emitted when a charged particle passes through a dielectric medium at a speed greater than the phase velocity of light in that medium.

Nuclear Magnetic Resonance

Nuclear magnetic resonance (NMR) is a technique used to determine the molecular structure of the substance by applying the external magnetic field to the atomic nuclei. It has wide applications in medicine, tool analysis, analyzing quality and purity of the sample etc.

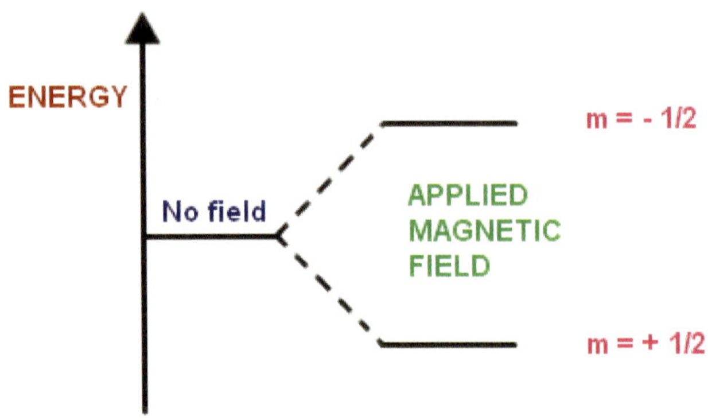

Energy Level for a Nucleus with spin Quantum Number 1/2

Radioactivity

Radioactivity is a phenomenon which is discover by a French scientist Henri Becquerel in 1896. He verified that uranium is a powerful radiations that is capable of blackening a photographic plate. Here uranium was discovered about hundred years as these radiations are not observed.

Natural radioactivity is the phenomenon of spontaneous and uncontrollable emission of radiations. This process which is considered as the disintegration are not stopped or slowed down by the use of and instrument are not observed by Henri Becquerel in discover of radioactivity.

Elementary Particles

Subatomic particles are small particles composed of nucleons or atoms. Elementary particles and composite particles are two type of sub atomicparticles. Elementary particles are those which are not measurable and are not made up of other particles and composite particles are made of some other particles also. Some of the Elementary particles are lepton, Hadron, Photon, graviton, quarks etc.

Nuclear Spin

The Nuclear spin is different from the electron spin. The nuclear spin represents the total angular momentum of the nucleus. It is represented by symbol I. The nucleus is although, composed of neutrons and protons but it acts as if it is a single entity which has intrinsic angular momentum. The Nuclear spin depends on the mass number, if the mass number is odd then the nucleus has half-integer spin like the electron while if the nucleus has even mass number then its spin will be integer spin.

Nuclear Stability

Nuclear stability determines how stable the nucleus is? we know that the Nucleus is made of neutrons and protons. What causes them to stick together? Why do the protons not repel each other?

Surprisingly, the mass of the nucleus is less than the sum of the masses of the individual protons and neutrons which makeup the nucleus. The lost mass (mass defect) has been changed into the energy necessary to bind the nucleus together. Suppose the mass defect is $\Delta\Delta$ m. Then the energy binding the nucleus together Δ E is given by:

$$\Delta E = \Delta m \times c^2$$

where C is the speed of light.

If is divided by the number of nucleons in the nucleus, then the binding energy per nucleon is obtained. The higher this value is the more stable is the nucleus.

Fermi Energy

The Fermi Energy is the energy of the highest level of quantum state which is occupied by the fermions (like electrons, protons or neutrons) at the absolute zero temperature.

The Fermi energy is useful in determining the thermal and electrical characteristics of the solids.

Shell Model

The shell model of the atom was first proposed by Niels Bohr in 1923.

The Shell Model is the model which describes the distribution of the energy level into different shells in an atom or in the nucleus of the atom. A shell is the energy level where particles of same energy exist. The shell model describes the arrangement of the elementary particles in the different shells of the atoms.

For example, the electrons are distributed in different shells based on their energy level and their direction of their rotation.

Millikan Oil Drop Model

The Millikan oil drop experiment is commonly known as Millikan experiment and was first performed by Robert Millikan and Harvey Fletcher in 1909. Robert Millikan was the professor of University of Chicago at that time. The experiment was performed by Professor Millikan with the useful inputs from Mr. Fletcher.

In this experiment a charged oil drop was suspended vertically between the two electrodes. Then the electric field is applied to such an extent that the free fall of the oil drop, under the influence of gravitational pull, was stopped. With the known electric field and other parameters Millikan was succeeded to calculate the charge on the drop. He published his findings in 1913 and was awarded Nobel prize for physics in 1923 for his work on finding the charge of the electron.

Fission

Fission is the process whereby a some matter divides and becomes multiple states.

Here in Nuclear physics, Fission process is used to split up a single atomic nuclei into two or more fragments.

Nuclear Fission

The Nuclear fission is defined as the process of splitting a heavy nucleus into two nuclei of smaller mass. Eg: Consider a Uranium Nuclei.

The process of fission is represented as follows:

$${}_{92}U^{235}+{}_{0}n^{1} \text{ --> } {}_{92}U^{236} \text{ --> } {}_{56}Ba^{141}+{}_{36}Kr^{92}+3{}_{0}n^{1}+ \text{Energy}$$

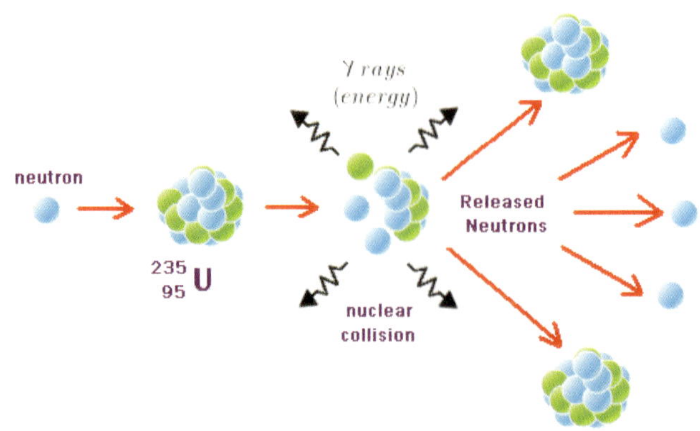

Nuclear Reactors

A Nuclear Reactor is used to extract the nuclear power from the radioactive material. The Nuclear reactor initiates and controls the nuclear chain reaction. The present day nuclear reactors are based on the Nuclear Fission Reaction.

Nuclear Structure

The atomic nucleus is a strongly-interacting, many-body quantum mechanical system that exhibits a fascinating variety of shapes and excitation modes, from spherical to super deformed (axis ratio 2:1), and from excitations of single protons and neutrons to collective vibrations and rotations of the nucleus as a whole. The study of nuclear structure attempts to elucidate the unifying mechanisms by which these rich patterns of behaviour emerge from the common underlying strong nuclear interaction between the nucleons (protons and neutrons) that form the nucleus.

Central to these studies is the concept of nuclear shell structure, in which protons and neutrons occupy quantized energy levels within a potential generated by their interactions with all of the other nucleons. The quantized energy levels cluster into groups, or shells, and the filling of major energy shells leads to particularly stable configurations associated with the proton and neutron "magic numbers": 2, 8, 20, 28, 50, 82, and 126. Nuclei near closed shells tend to have spherical ground states, while those far from shell closures are deformed and exhibit low-energy collective rotational excitations.

While the nuclear shell structure and magic numbers have been well studied close to the "valley of stability" defined by the ~ 300 stable isotopes found in nature, investigations of exotic nuclei with unusual proton-to-neutron ratios are revealing an evolution of this shell structure as one moves away from the stable isotopes. Understanding of this complex evolution of nuclear shell structure is of fundamental interest both as an emergent property of the nuclear quantum many-body system and as an important input to understanding the nuclear reaction processes by which the heavy chemical elements have been, and continue to be, synthesized in explosive astrophysical environments. Nuclear shell structure in extremely neutron-rich nuclei plays a major role in the pathway for the rapid neutron capture (r) process, thought to be responsible for producing approximately half of the elements heavier than iron. Neither the exact locations of the shell gaps in these neutron-rich isotopes, nor the exact astrophysical events in which the r-process takes place, are, however, definitively established at present.

Gamma-ray spectroscopy provides powerful techniques to study nuclear shell structure through measurements of gamma-ray energies and angular correlations in low-energy decay experiments and through measurements of electromagnetic transition matrix elements between nuclear states in Coulomb excitation experiments with accelerated radioactive ion beams.

Models

The Liquid Drop Model

The liquid drop model is one of the first models of nuclear structure, proposed by Carl Friedrich von Weizsäcker in 1935. It describes the nucleus as a semiclassical fluid made up of neutrons and protons, with an internal repulsive electrostatic force proportional to the number of protons. The quantum mechanical nature of these particles appears via the Pauli exclusion principle, which states that no two nucleons of the same kind can be at the same state. Thus the fluid is actually what is known as a Fermi liquid. In this model, the binding energy of a nucleus with Z protons and N neutrons is given by,

$$E_B = a_V A - a_S A^{2/3} - a_C \frac{Z^2}{A^{1/3}} - a_A \frac{(N-Z)^2}{A} - \delta(A,Z)$$

where $A = Z + N$ is the total number of nucleons (Mass Number). The terms proportional to A and $A^{2/3}$ represent the volume and surface energy of the liquid drop, the term proportional to Z^2 represents the electrostatic energy, the term proportional to $(N-Z)^2$ represents the Pauli exclusion principle and the last term $\delta(A,Z)$ is the pairing term, which lowers the energy for even numbers of protons or neutrons. The coefficients a_V, a_S, a_C, a_A and the strength of the pairing term may be estimated theoretically, or fit to data. This simple model reproduces the main features of the binding energy of nuclei.

The assumption of nucleus as a drop of Fermi liquid is still widely used in the form of Finite Range Droplet Model (FRDM), due to the possible good reproduction of nuclear binding energy on the whole chart, with the necessary accuracy for predictions of unknown nuclei.

There are many models to study the nucleus structure. Liquid Drop Model was the first model proposed. Later came the Nuclear Shell Modelwhich was first proposed by Maria Goeppert-Mayer in 1948.

The Nuclear Shell Model deals with the features of energy levels. A shell is the energy level where particles of same energy exist. The Shell Model describes the arrangement of the elementary particles in the different shells of the atoms.

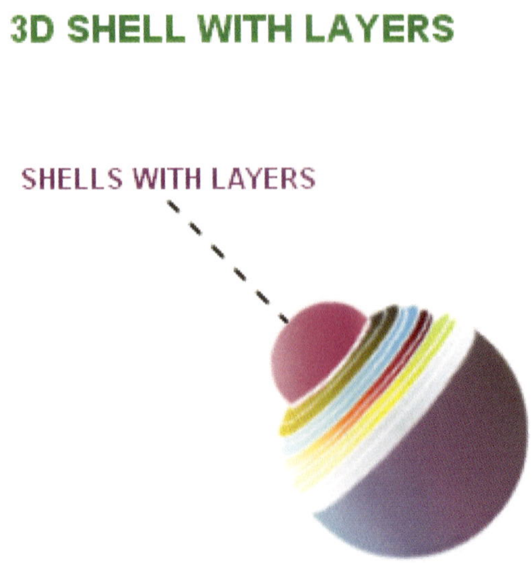

An electron shell may be thought of as an orbit followed by electrons around an atom nucleus. Because each shell can contain only a fixed number of electrons, each shell is associated with a particular range of electron energy and thus each shell must fill completely before electrons can be added to an outer shell. The electrons in the outermost shell determine the chemical properties of the atom.

A Shell is the energy level where particles of same energy exist. The Shell Model is the model which describes the distribution of the energy level into different shells in an atom or in the nucleus of the atom.

Valance shell electronic configuration is the representation of the participation of electrons in orbits of atom. The electrons in an atom are filled by the three basic principles:

1. AufBau Principle
2. Pauli Exclusion Principle
3. Hund's Rule

As per AufBau principle, the electrons will be filled first in orbitals with lower energy like 1s will be filled first before 2s and 2s will be filled before 2p. In similar way the electrons will get filled in lower energy level then go for higher energy level.

As per Pauli's exclusion principle, no two electrons within a particular atom can have identical quantum numbers. In similar meaning it says that two electrons occupy the same orbital must have opposite spins.

As per Hund's rule, when an electron come to an atom and has to choose between two or more orbitals of same energy, then the preference will be given to empty orbital rather than the occupied one. As more electrons will be occupied, they tend to follow half-filled and fully filled rule.

The electron shells are labelled K, L, M, N, O, P, and Q or 1, 2, 3, 4, 5, 6, and 7 going from innermost shell outwards.

The order of filling of the electrons in an atom will be like this:

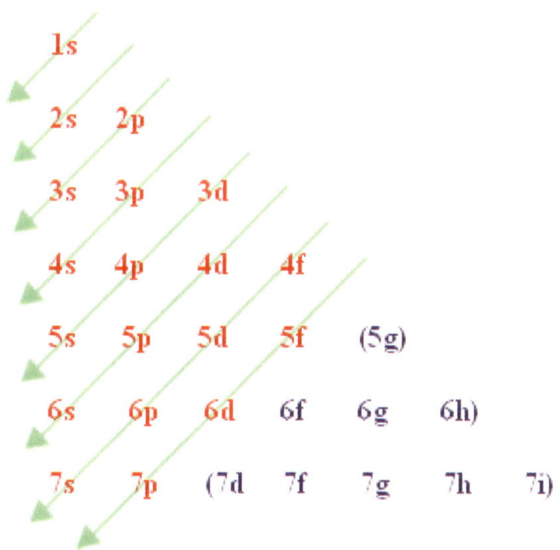

The orbits of electrons are as below:

1s, 2s, 2p, 3s, 3p, 4s, 3d, 4p, 5s, 4d, 5p, 6s, 4f, 5d, 6p, 7s, 5f, 6d, 7p.

Atom reacts Chemically and still behaves as a Conductor

Electrons in outer shells have higher average energy and travel farther from the nucleus than those in inner shells. Here the pull of the atom's nucleus upon them is weaker and more easily broken.

Shell Model of the Atom

The Model of Hydrogen was extended by Neil Bohr to give a rough model outlook to the heavier atoms. This approximation gave a physical picture which on reproduction throws light on many physical properties for the first time in the history of chemical sciences.

It was postulated that heavier atoms had higher number of protons in the nucleus and also had equivalent high level of electrons to compensate the charges.

Bohr postulated that:

> "Each single orbit had the capacity to hold only a specific number of electrons. Once the specific number of electron is filled up the next orbital has to be used for filling new set of electrons".

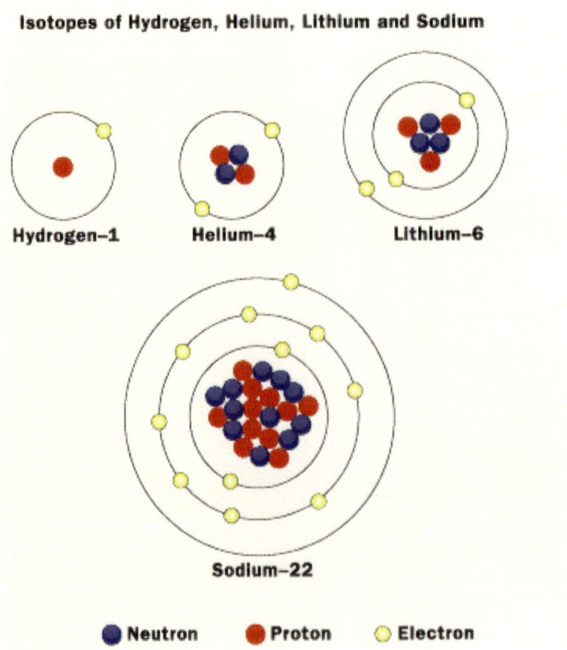

The electrons are distributed in different shells based on their energy level and the direction of their rotation. Besides electrons, the nucleus is also said to be divided into several shells and the protons and neutrons are distributed in these shells as opposed to the previous concept that the protons and neutrons are just concentrated in the center which is nucleus of the atom.

The shell model of the atom is the configuration of the electrons in the atom. The shell is the set of the allowed states of the electrons which shares common principle quantum number n. The nth shell of an atom consists of $2n^2$ electrons.

The first shell, for example consists of 2 electrons while second shell consists of 8 electrons. A shell can further be divided into sub shells.

A Sub shell is the set of states of the electrons which shares common azimuthal quantum number. The azimuthal quantum number determines the orbital angular momentum and also the shape of the orbit.

The maximum number of the electrons placed in the sub shells are given by 2(2 u +1). The u values of 1, 2, 3, 4 are corresponds to s, p, d, f labels of the shell.

Nuclear Shell Model

The Shell model of the nucleus was first proposed by Maria Goeppert Mayer in 1948. She is the one of the only two women who has received Nobel Prize for physics for her path breaking discovery.

In this Model, the nuclei of atoms are also divided into different shells and the nucleons (protons and neutrons) are distributed into these shells in the same manner that the electrons are distributed in the different shells of the atoms.

With this model of nuclear shells, she was able to justify why few elements are stable and few are not. With this nuclear shell model the stability of the certain nucleus are justified.

More importantly it is observed that the nuclei associated with the "magic numbers" of protons are more stable with respect to other nuclei. The magic numbers are : 2,8,20,28,50,82,126,184.

A nucleus with the magic number both of protons and neutron are known as Doubly Magic Nuclei. This model also signifies why isotopes of certain elements (for e.g. calcium) are all stable while the isotopes of few isotopes of other are not all stable.

Electron Shell Model

The Electron Shell is an orbit which is followed by electrons around an atom's nucleus. The shells are described with respect to the numerals. The most nearest shell to the nucleus is designated as '1' shell. The shell contains only fixed number of the electrons and the number is governed by principle quantum number n as $2n^2$ electrons. These shells are divided into sub shells which are governed by azimuthal quantum number.

The electron shell are represented either by numeral or by alphabets such as K, L, M, N, O, P, Q going from innermost shell to outwards. The outermost shell contains electrons with higher average energy.

The various possible sub shells of electron or atom shell model are shown below:

Subshell	ι	Max electrons	Shell containing it	Historical Name
s	0	2	Every Shell	Sharp
p	1	6	2nd and higher shell	principal
d	2	10	3rd and higher shell	diffuse
f	3	14	4th and higher shell	fundamental
g	4	18	5th and higher shell	(next in alphabet after f)

In the above table the second column gives the characteristics of the sub shell and the maximum number of electron in that sub shell is dependent on this factor. The maximum number of electrons in the sub shell is governed by 2 (2 ι+1).

Shell Structure of Atom and Nuclei

Although the basic concept behind both the shell structure is same, that is the arrangements of the fundamental particles in the different energy shells, but there are certain difference as regard to both the system.

Below is the diagram showing shell structure of both the atom as well as nuclei:

S.No	Shell structure of Nuclei	Shell structure of atom
1	Here Energy state starts from Zeroth state.	Here Energy state starts from first state.
2	The Magic numbers are : Z = 2, 8, 20, 28, 50, 82,126 and so on.	The Magic numbers are : Z = 2,10,18, 36, 54, 86 and so on.
3	The average potential of shell structure of Nuclei is different from that of atom	The average Potential varies.
4	Here the particles are neutrons and protons	Here the particles are electrons.

Although the shells are divided into sub shells, these sub shells themselves again sub divided into multiplicity of sub shells according to the spin of the electrons (atomic or electronic model) or the nucleons (Nuclei model).

Mean Field Theories

Independent-particle Model

The interaction between nucleons, which is a consequence of strong interactions and binds the nucleons within the nucleus, exhibits the peculiar behaviour of having a finite range: it vanishes when the distance between two nucleons becomes too large; it is attractive at medium range, and repulsive at very small range. This last property correlates with the Pauli exclusion principle according to which two fermions (nucleons are fermions) cannot be in the same quantum state. This results in a very large mean free path predicted for a nucleon within the nucleus.

The main idea of the Independent Particle approach is that a nucleon moves inside a certain potential well (which keeps it bound to the nucleus) independently from the other nucleons. This amounts to replacing a *N*-body problem (*N* particles interacting) by *N* single-body problems. This essential simplification of the problem is the cornerstone of mean field theories. These are also widely used in atomic physics, where electrons move in a mean field due to the central nucleus and the electron cloud itself.

The independent particle model and mean field theories (we shall see that there exist several variants) have a great success in describing the properties of the nucleus starting from an effective interaction or an effective potential, thus are a basic part of atomic nucleus theory. One should also notice that they are modular enough, in that it is quite easy to extend the model to introduce effects such as nuclear pairing, or collective motions of the nucleon like rotation, or vibration, adding the corresponding energy terms in the formalism. This implies that in many representations, the mean field is only a starting point for a more complete description which introduces correlations reproducing properties like collective excitations and nucleon transfer.

Nuclear Potential and Effective Interaction

A large part of the practical difficulties met in mean field theories is the definition (or calculation) of the potential of the mean field itself. One can very roughly distinguish between two approaches:

- The phenomenological approach is a parameterization of the nuclear potential by an appropriate mathematical function. Historically, this procedure was applied with the greatest success by Sven Gösta Nilsson, who used as a potential a (deformed) harmonic oscillator potential. The most recent parameterizations are based on more realistic functions, which account more accurately for scattering experiments, for example. In particular the form known as the Woods–Saxon potential can be mentioned.
- The self-consistent or Hartree–Fock approach aims to deduce mathematically the nuclear potential from an effective nucleon–nucleon interaction. This technique implies a resolution of the Schrödinger equation in an iterative fashion, starting from an ansatz wavefunction and improving it variationally, since the potential depends there upon the wavefunctions to be determined. The latter are written as Slater determinants.

In the case of the Hartree–Fock approaches, the trouble is not to find the mathematical function which describes best the nuclear potential, but that which describes best the nucleon–nucleon interaction. Indeed, in contrast with atomic physics where the interaction is known (it is the Coulomb interaction), the nucleon–nucleon interaction within the nucleus is not known analytically.

There are two main reasons for this fact. First, the strong interaction acts essentially among the quarks forming the nucleons. The nucleon–nucleon interaction in vacuum is a mere *consequence* of the quark–quark interaction. While the latter is well understood in the framework of the Standard Model at high energies, it is much more complicated in low energies due to color confinement and asymptotic freedom. Thus there is yet no fundamental theory allowing one to deduce the nucleon–nucleon interaction from the quark–quark interaction. Furthermore, even if this problem were solved, there would remain a large difference between the ideal (and conceptually simpler) case of two nucleons interacting in vacuum, and that of these nucleons interacting in the nuclear matter. To go further, it was necessary to invent the concept of effective interaction. The latter is basically a mathematical function with several arbitrary parameters, which are adjusted to agree with experimental data.

Most modern interaction are zero-range so they act only when the two nucleons are in contact, as introduced by Tony Skyrme.

Self-consistent Approaches of the Hartree–Fock Type

In the Hartree–Fock approach of the n-body problem, the starting point is a Hamiltonian containing n kinetic energy terms, and potential terms. As mentioned before, one of the mean field theory hypotheses is that only the two-body interaction is to be taken into account. The potential term of the Hamiltonian represents all possible two-body interactions in the set of n fermions. It is the first hypothesis.

The second step consists in assuming that the wavefunction of the system can be written as a Slater determinant of one-particle spin-orbitals. This statement is the mathematical translation of the independent-particle model. This is the second hypothesis.

There remains now to determine the components of this Slater determinant, that is, the individual wavefunctions of the nucleons. To this end, it is assumed that the total wavefunction (the Slater determinant) is such that the energy is minimum. This is the third hypothesis.

Technically, it means that one must compute the mean value of the (known) two-body Hamiltonian on the (unknown) Slater determinant, and impose that its mathematical variationvanishes. This leads to a set of equations where the unknowns are the individual wavefunctions: the Hartree–Fock equations. Solving these equations gives the wavefunctions and individual energy levels of nucleons, and so the total energy of the nucleus and its wavefunction.

This short account of the Hartree–Fock method explains why it is called also the variational approach. At the beginning of the calculation, the total energy is a "function of the individual wavefunctions" (a so-called functional), and everything is then made in order to optimize the choice of these wavefunctions so that the functional has a minimum – hopefully absolute, and not only local. To be more precise, there should be mentioned that the energy is a functional of the density, defined as the sum of the individual squared wavefunctions. Let us note also that the Hartree–Fock method is also used in atomic physics and condensed matter physics as Density Functional Theory, DFT.

The process of solving the Hartree–Fock equations can only be iterative, since these are in fact a Schrödinger equation in which the potential depends on the density, that is, precisely on the wavefunctions to be determined. Practically, the algorithm is started with a set of individual grossly reasonable wavefunctions (in general the eigenfunctions of a harmonic oscillator). These allow to compute the density, and therefrom the Hartree–Fock potential. Once this done, the Schrödinger equation is solved anew, and so on. The calculation stops – convergence is reached – when the difference among wavefunctions, or energy levels, for two successive iterations is less than a fixed value. Then the mean field potential is completely determined, and the Hartree–Fock equations become standard Schrödinger equations. The corresponding Hamiltonian is then called the Hartree–Fock Hamiltonian.

Relativistic Mean Field Approaches

Born first in the 1970s with the works of John Dirk Walecka on quantum hadrodynamics, the relativistic models of the nucleus were sharpened up towards the end of the 1980s by P. Ring and coworkers. The starting point of these approaches is the relativistic quantum field theory. In this context, the nucleon interactions occur via the exchange of virtual particlescalled mesons. The idea is, in a first step, to build a Lagrangian containing these interaction terms. Second, by an application of the least action principle, one gets a set of equations of motion. The real particles (here the nucleons) obey the Dirac equation, whilst the virtual ones (here the mesons) obey the Klein–Gordon equations.

In view of the non-perturbative nature of strong interaction, and also in view of the fact that the exact potential form of this interaction between groups of nucleons is relatively badly known, the use of such an approach in the case of atomic nuclei requires drastic approximations. The main simplification consists in replacing in the equations all field terms (which are operators in the mathematical sense) by their mean value (which are functions). In this way, one gets a system of coupled integro-differential equations, which can be solved numerically, if not analytically.

Interacting Boson Model

The interacting boson model (IBM) is a model in nuclear physics in which nucleons are represented as pairs, each of them acting as a boson particle, with integral spin of 0, 2 or 4. This makes calculations feasible for larger nuclei. There are several branches of this model - in one of them (IBM-1) one can group all types of nucleons in pairs, in others (for instance - IBM-2) one considers protons and neutrons in pairs separately.

Spontaneous Breaking of Symmetry in Nuclear Physics

One of the focal points of all physics is symmetry. The nucleon–nucleon interaction and all effective interactions used in practice have certain symmetries. They are invariant by translation (changing the frame of reference so that directions are not altered), by rotation (turning the frame of reference around some axis), or parity (changing the sense of axes) in the sense that the interaction does not change under any of these operations. Nevertheless, in the Hartree–Fock approach, solutions which are not invariant under such a symmetry can appear. One speaks then of spontaneous symmetry breaking.

Qualitatively, these spontaneous symmetry breakings can be explained in the following way: in the mean field theory, the nucleus is described as a set of independent particles. Most additional correlations among nucleons which do not enter the mean field are neglected. They can appear however by a breaking of the symmetry of the mean field Hamiltonian, which is only approximate. If the density used to start the iterations of the Hartree–Fock process breaks certain symmetries, the final Hartree–Fock Hamiltonian may break these symmetries, if it is advantageous to keep these broken from the point of view of the total energy.

It may also converge towards a symmetric solution. In any case, if the final solution breaks the symmetry, for example, the rotational symmetry, so that the nucleus appears not to be spherical, but elliptic, all configurations deduced from this deformed nucleus by a rotation are just as good solutions for the Hartree–Fock problem. The ground state of the nucleus is then *degenerate.*

A similar phenomenon happens with the nuclear pairing, which violates the conservation of the number of baryons.

Extensions of the Mean Field Theories

Nuclear Pairing Phenomenon

The most common extension to mean field theory is the nuclear pairing. Nuclei with an even number of nucleons are systematically more bound than those with an odd one. This implies that each nucleon binds with another one to form a pair, consequently the system cannot be described as independent particles subjected to a common mean field. When the nucleus has an even number of protons and neutrons, each one of them finds a partner. To excite such a system, one must at least use such an energy as to break a pair. Conversely, in the case of odd number of protons or neutrons, there exists an unpaired nucleon, which needs less energy to be excited.

This phenomenon is closely analogous to that of Type 1 superconductivity in solid state physics. The first theoretical description of nuclear pairing was proposed at the end of the 1950s by Aage Bohr, Ben Mottelson, and David Pines (which contributed to the reception of the Nobel Prize in Physics in 1975 by Bohr and Mottelson). It was close to the BCS theory of Bardeen, Cooper and Schrieffer, which accounts for metal superconductivity. Theoretically, the pairing phenomenon as described by the BCS theory combines with the mean field theory: nucleons are both subject to the mean field potential and to the pairing interaction.

The Hartree–Fock–Bogolyubov (HFB) method is a more sophisticated approach, enabling one to consider the pairing and mean field interactions consistently on equal footing. HFB is now the de facto standard in the mean field treatment of nuclear systems.

Symmetry Rrestoration

Peculiarity of mean field methods is the calculation of nuclear property by explicit symmetry breaking. The calculation of the mean field with self-consistent methods (e.g. Hartree-Fock), breaks rotational symmetry, and the calculation of pairing property breaks particle-number.

Several techniques for symmetry restoration by projecting on good quantum numbers have been developed.

Particle Vibration Coupling

Mean field methods (eventually considering symmetry restoration) are a good approximation for the ground state of the system, even postulating a system of independent particles. Higher-order corrections consider the fact that the particles interact together by the means of correlation. These correlations can be introduced taking into account the coupling of independent particle degrees of freedom, low-energy collective excitation of systems with even number of protons and neutrons.

In these way excited states can be reproduced by the means of random phase approximation (RPA), and eventually consistently calculating also corrections to the ground state (e.g. by the means of nuclear field theory).

Nuclear Magnetic Resonance

Nuclear Magnetic Resonance (NMR) is a technique is based on the Principle that when the hydrogen atoms (of water molecules) are subjected to a very strong magnetic field and radio waves, hydrogen atoms release protons. It was first developed by Felix Bloch and Purcell in USA (1946). They shared the Nobel Prize for Physics in 1952.

This technique is also called Magnetic Resonance Imaging. It has its wide applications in analyzing molecular structure and also for determining purity of the sample.

NMR is also called Nuclear Magnetic Resonance. It is a physical phenomenon. When external magnetic field in applied magnetic nuclei absorb and re-emits radiations that are electromagnetic. This energy that is present is actually at a specific resonance frequency.

All Nuclei are electrically charged in nature. If an external magnetic field (B_0) is applied to an Hydrogen atom, the nuclei spin will align with the field.

There are only two orientations that the spin $\frac{1}{2}$ nuclei can adopt:

- Either aligned with applied Magnetic field (parallel or spin + $\frac{1}{2}$)

or

- Aligned against the field (Antiparallel or spin - $\frac{1}{2}$).

There is a random orientation of the magnetic moments or an orientation of equal energy when there is no magnetic field. There is an opposite alignment of spins in the presence of external magnetic field.

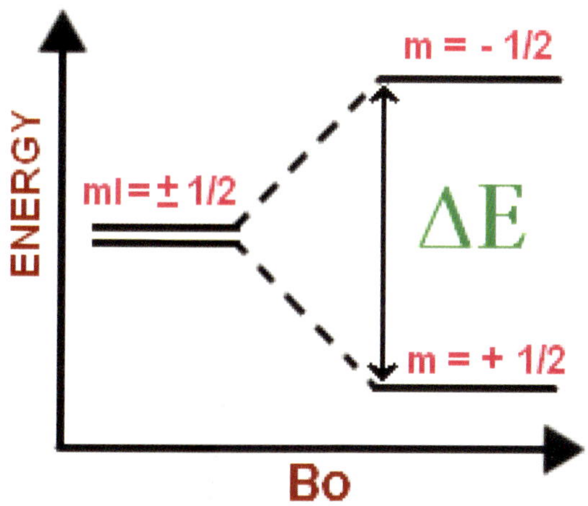

Working

If the magnetic field is applied then the energy levels would be splitted. New magnetic quantum number will be obtained from each splitted level.

When energy is absorbed the nuclei excites from the lower spin state to the upper spin state. As a result of excitation the signal of the NMR spectrum is produced.

The strength of external magnetic field decides the difference in energy that exists between the two spin states.

After some time the spin returns to the base level emitting the energy at same frequency. The signal matches this transfer which gives the NMR spectrum.

The graphical representation between absorption and the applied radio frequency is called NMR spectrum.

Nuclear Magnetic Moment

With each nuclear spin a magnetic moment is associated. Magnetic moments can be expressed in terms of the nuclear spin.

This is a manner similar to the treatment of the magnetic moments of electron spin and electron orbital angular momentum. Bohr Magneton is used for expressing the magnetic moment.

Nuclide	Nuclear spin	Magnetic Moment
n	1212	-1.92
p	1212	+2.8
2H (D)	1	+0.86
17 O	5252	-1.9
57 Fe	1212	+0.091
57 CO	7272	+4.73
93 Nb	9292	+6.2

Nuclear Magnetic Resonance Spectroscopy

The interaction of energy with a sample for the analysis of that specific sample is done by Spectroscopy Technique. From spectroscopy we obtain data which is called spectrum.

Nuclear Magnetic Resonance Spectroscopy is basically a graph of the detected intensity of energy against the wavelength, momentum or frequency.

This spectrum has various uses like knowing the information about molecular geometry, bonds etc.

There are many types of spectroscopic techniques such as NMR, mass, IR, UV or visible etc.

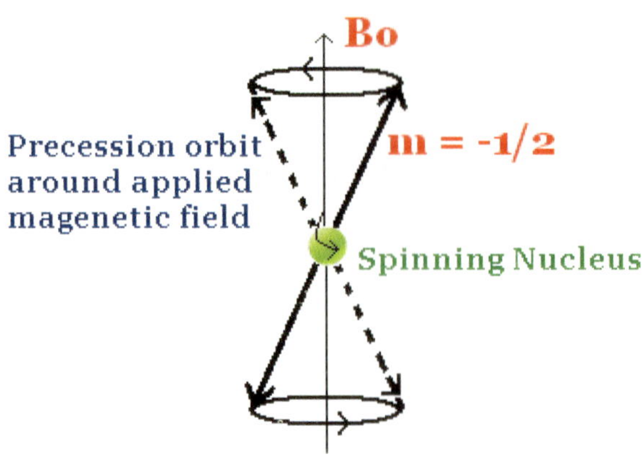

Principle of NMR

The Principle of NMR is that the nucleus of the atom which has odd mass and atomic number contains both the spin angular momentum and a magnetic moment.

The notation for the possible orientated spin states are 2I + 1.

Here, I = Nuclear spin quantum number.

The frequency of precession of the spin is proportional to the strength of the magnetic field, as noted by the equation:

$$\omega o = \gamma B_o$$

The frequency ω0ω0 is called the Larmor frequency and is given by radians per second. The proportionality constant γγ is the gyromagnetic ratio.

The basic requirement of NMR spectroscopy is strong magnetic field. The structure of molecules can be found by NMR spectroscopy.

Note: The external magnetic field and frequency of radiation are the factors on which NMR depends.

Solid State Nuclear Magnetic Resonance

It is a kind of nuclear magnetic resonance (NMR) spectroscopy. It is characterized by the presence of anisotropic interactions. Anisotropic interactions are basically directionally dependent interactions.

For example: Magnetic Anisotropy is basically the non uniform magnetic field. The existence of electron system in the compounds like aromatics, alkenes etc. leads to its existence.

During the application of magnetic field the electron system will interact with it thus generating an induced magnetic field which is the basic reason for the anisotropy.

There are many examples of chemical anisotropic nuclear interaction like CSA and internuclear bipolar coupling. The J coupling is yet another example.

The nuclear spin energy levels are altered as a result of anisotropic interactions.

There is a line broadening effect as a result of this in the spectra of NMR. It has application in chemistry and biology.

Applications

The applications of NMR are in various fields like:

- Medicines: in MRI etc.
- Chemistry: to determine the structure of various compounds.
- In Non destructive testing.
- In the acquisition of dynamic information.
- For data acquisition in petroleum industry.
- In flow probes in NMR spectroscopy.
- In process control.
- For determining earth's filed NMR.
- For quantum computing.
- For making magnetometers.

The NMR technique is used to detect the atomic resolution detection of biomacromolecules and native membrane protein. It is done in aqueous solution in the physiological or membrane mimetic environments. The solid state NMR technique is used for that.

Radioactivity

The Radioactivity discovery has got a quite amazing history behind it.

In 1896, Henri Becquerel observed that when a photographic plate wrapped in black paper was placed near a salt of uranium it got affected having bright spots on it. He was surprised to see this as he had wrapped the plates in black paper. So he conducted the same experiment with other salts of uranium. This led him to conclude that the uranium salt emitted some deep penetrating radiation.

Further by conducting more experiments it was realized that the intensity of the emitted radiation depended directly on the concentration of uranium in its various salts. The emission of radiation was completely unaffected by any change in the physical and chemical conditions of the system. Becquerel concluded that the origin of the radiation was somehow rooted in the nucleus of the uranium atom. The radiation discovered by Becquerel was initially named Becquerel rays.

Later on Madam Curie and Pierre Curie discovered other substances i.e., polonium, radium etc., which were more active than uranium. These substances, which have the property of spontaneous emission of radiation are called radioactive substances and the process of spontaneous emission of radiation is called Radioactivity.

Radioactivity is the phenomena in which radioactive elements like uranium, polonium and radium undergoes decay emitting radiations.

Consider unstable nuclei like Uranium, when it undergoes radioactive decay, it decays to form Thorium emitting radiations.

$$^{238}U_{92} \text{ --> } ^{234}Th_{90} + {}^{4}He_{2} + 2\ {}^{0}\gamma\gamma_{0}$$

Here Uranium is the Parent nuclei and new element produced in the disintegration Thorium is the daughter nuclei.

Radioactivity is defined as the self spontaneous disintegration that is decay of unstable radioactive nuclei and the nucleus exhibiting this phenomenon are known as radioactive nuclei. Some examples of radioactive substances are Uranium, Radium, Thorium, Polonium, Neptunium etc.

When the process of radioactivity takes place, the nuclei undergoes the disintegration as per the Laws of radioactive disintegration which states that,

If N = Number of active nuclei at time t,

N – dN = Number of active nuclei after time interval of dt.

The Rate of radioactive decay is given by:

$$\text{Rate of decay} = -\frac{dN}{dt}$$

According to the Rutherford and Soddy law of radioactive decay, at any instant the rate of decay of a radioactive material is directly proportional to the number of atoms present at that instant. Hence rate of decay is directly proportional to N.

If N = No of active nuclei left at time t.

$N = N_0^{e-\lambda t}$ The atoms which possess radioactivity or whose disintegration releases radioactive rays are called Radioactive Atoms.

Causes

Radioactivity occurs in unstable nuclei. It is realized only when there is excess nucleons present in the nucleus. if excess mass is present in the nuclei, it emits alpha particle. If more neutrons are there, it emits beta particle and if excess protons are there, it emits positron while excess energy leads to the emission of gamma rays.Decay processes are random. Here one simply knows that in a radioactive element, radioactivity is taking place or it is definite that a certain number of atoms will decay in a given time interval but never knows that which particular radioactive nuclei will decay when. It is just a matter of chance that is probability which is explained by quantum mechanics and it is statistical in nature.

Radioactivity is a nuclear event and not atomic and hence electronic configuration of atom doesn't have any relationship with radioactivity.

Radioactive Half Life

Decay half life was used to describe the characteristic of unstable atoms, that is radioactive decay. It was given by the famous scientist Antonie Henri Becquerel in 1896. It is defined as:

> "The time taken by the substance undergoing decay to decrease by half of its original amount."

According to law of radioactivity, we know that,

$$\frac{N}{N_0}\left(\frac{1}{2}\right)^{\left(\frac{t}{T_{1/2}}\right)}$$

Where N = No of active nuclei at time t,

N_0 = Initial amount of radioactive substances at time t = 0,

$t_{1/2}$ = Half life of radioactive substances,

t = Time taken by the radioactive substance to disintegrate.

The number of radioactive elements present is given by soddy displacement law as:

$N = N_0^{e-\lambda t}$

Where λλ = Decay Constant.

When we consider the half life $t=T_{\frac{1}{2}}$ (half life) then $N = \frac{N_0}{2}$

Taking logarithm on both the sides we get,

$$\lambda T_{\frac{1}{2}} = \log_e 2$$

$$\lambda T_{\frac{1}{2}} = 2.303 \log_{10} 2$$

$$T_{\frac{1}{2}} = \frac{2.303 \log_{10} 2}{\lambda}$$

$$T_{\frac{1}{2}} = \frac{0.693}{\lambda}$$

Thus the decay half life of the radioactive substance is inversely proportional to the decay constant.

Mean life of a radioactive nuclei is equal to that time in which number of nuclei left becomes either $\frac{1}{e}$ times the original nuclei or approximately 37% of the original nuclei.

Number of decayed nuclei in mean life = $N_0 - N$

$$= N_0 - \frac{N_0}{e}$$

$$= N_0\left(1 - \frac{1}{e}\right)$$

$$= 0.63N_0$$

Mean lift is equal to that time in which number of decayed nuclei becomes either $\frac{e-1}{e}$ times original nuclei or approximately 63 % of original nuclei.

Radioactive Waste

Radioactive waste is one which is generated during the disposal of nuclear related weapons and during the production of electricity by nuclear power.The majority of Radio active waste present is of "Low - Level Radio Waste" that means it has lower level of radio activity with respect to mass.

Radioactive wastes are categorized into three types:

1. Transuranic waste
2. Low-level waste
3. High-level waste.

Transuranic Waste

It is comprised of waste that is produced from nuclear weapons. Transuranic waste is a long-life waste which is also called as Trans-uranic waste, because it comprise of elements which is much heavier than uranium. These type of wastes are contaminated with alpha-particles which is produced during weapon production.

Although Transuranic radio waste is not much effective as it does not produce high-level penetrating radiation but if the alpha-particles that are associated with Transuranic waste is inhaled then it will damage lungs and some of the internal organs.

Low-level Waste

It occupies the majority of radioactive waste as it comprises clothing, tools, paper, rags and so on. This type of waste is generated from nuclear power plants, hospitals and industries.

Low-level contamination takes place through isotopes which is of short half-life or long-life, a minimal precaution is enough for short half -life But care should be taken in case of Longer-lived isotopes and it should be disposed and cleaned them properly, because even the low level contamination waste may also affects the human life when he come across for exposure for longer time.

High Level Waste

It is a type of waste generated from nuclear power reactors. High level radioactive waste is dangerous and it requires special precautions from human beings and also requires a special care on remote handling.

High-level radioactive waste has a great application as it is used to produce at most 1,000,000kW in one year at the same time if the this radioactive waste is not used in the right way it is great threat to man kind because the waste produced by this radioactive waste has a capacity to kill almost 10 billion peoples.

Radioactive Pollution

Radioactive pollution is caused by harmful ionizing radiations. Atom bomb explosion, nuclear power plant accident etc are nuclear hazards. It is a radioactive pollution. It is caused by radioactive substances called radioactive isotopes or radionuclides. There is a Well known living example for this: Radioactive pollution causes holocaust. It is the mass extermination of human beings by fire. It is a nuclear hazard. In Hiroshima and Nagasaki, 3 lakhs and 40,000 people were killed during these two atom bomb explosion. Atom bomb explosion liberates enormous heat due to atomic fission reactions. This heat melts even stones and metals. Hence man, animals and plants are burnt to death. The Chernobyl atomic power plant accident caused the death of 2000 people.It can causes harmful, irrepairable effect to the life kind. It is also injurious to health and even causes death. Some are given as below:

1. Radiations cause gene mutations and chromosomal aberrations. This leads to genetical defects in future generation.
2. It can cause Stillbirths - Birth of dead babies, Congenital deformities - Defects in new born babies, Microcephaly - Abnormal small head in babies.
3. Ionizing radiations penetrate into the body and cause ionization of molecules in cells. These cells cannot function normally and die.

4. It causes radiation sickness. It is characterized by reddened and alcerated skin, loss of hair, nausea, anaemia, etc.

Radioactive Contamination

Due to the nuclear power plants and Nuclear energy industry there are a lot of contamination occurring in the environment. The wastes of radioactive element are emitting the radiations which are contaminating the soil, air as well as water.

- Nuclear contamination in water: Water bodies like lake, wells, rivers, ponds and other bodies occupy almost 70% of the earth surface. So water pollution due to the nuclear wastes will affect the marine animals badly. This kind of water if consumed by the human being will lead him to a series of dreadful diseases like cholera, typhoid, Ascariasis, Jaundice, gasteroenteritis, Giardiasis etc.The effect of water pollution on growing foetus is very dangerous in pregnant women. It causes delayed or incomplete mental development, autism or brain damage of foetus.
- Nuclear contamination in air: Harmful gases given out by the industries and power plants get mixed with water vapor, which form clouds and fall down as acid rain. Acid rains containing sulphates is another cause of water pollution that affects plants and animals. Acid rains are harmful for fish and marine life in aquatic environment.
- Nuclear contamination in soil: There are two kinds of wastes that are generated from industrial activities, unused chemicals and unwanted industrial garbage. These wastes mix with soil and affects a lot.

  Unused chemicals like fly ash, sludge, plastics and saw dust and some of the chemicals that are discharged from the industries.The spoil the fertility of the soil. The chemicals released are harmful to living organisms dependent on the soil. Contamination of the soil is a direct cause of contamination of the crops.As a result harmful chemicals enter the food chain.

Radioactive Exposure

Radioactive exposure has some serious consequences. Some of them are:

1. The most serious effect of nuclear pollution is on the humans and animals. The harmful radiations are the reason of some disease like cancer. The radiation destroys the enzymes of human body. Different organs of human body show different sensitivity towards radiations. The short range effect shows the immediate effect which generally shows with in two or three days. The short range effects involves loss of hair or nails, subcutaneous bleeding, and change in the number and proportion of cells and metabolism.
2. The long range effects or delayed effects take few months or some years to occur. They involve the changes in genetic order, mutations, tumors, and alteration of life span.
3. It also affects the animals and plants. The radioactive materials are transferred in the body of plants and animals by the food chain in the form of zinc (Zn), iron (Fe) etc.

4. The working employees in the uranium mines are suffered from skin burn and cancer type serious diseases due to harmful radioactive material.

Radioactive Substances

Uranium, thorium, nuclear power plant left overs, coolant used etc are all substances containing radioactive material. These in true sense are called Radioactive Substances.

Uses of Radioactive Isotopes

In the Field of Medicine

1. For testing blood cr-51
2. For testing blood circulation – Na -24
3. For detecting brain tumor – radio mercury -203
4. For detecting fault in thyroid gland – radio iodine – 131
5. For cancer – co-60
6. For blood - Au-189
7. For skin diseases – P- 31.

In the Field of Archaeology

1. For determining age of archaeological sample C-14
2. For determining age of meteorites – K- 40
3. For determining age of earth – lead isotopes.

In the Field of Agriculture

1. For protecting potato crop from earthworm – CO- 60
2. For artificial rains – AgI
3. As fertilizers – P- 32.

As Tracers

A very small amount of radioisotopes present in a mixture is known as a tracer.

It is used for studying biochemical reaction in tracer and animals.

In Industries

It is used for detecting leakages in oil or water pipe lines, and for determining the age of planets.

Nuclear Spin

Many atomic nuclei have a property called spin: the nuclei behaves as if they were spinning. In fact, any atomic nucleus that possesses either odd mass, odd atomic number, or both has a quantized spin angular momentum and a magnetic moment. The important consequence of nuclear spin is the energy splitting of degenerate Nuclear Spin states under an external magnetic field. It is possible to drive transition between different spin states by applying radio-frequency electromagnetic radiation. In this section we will learn more about nuclear spin.

The Nuclear Spin is different from the electron spin. The nuclear spin represents the total angular momentum of the nucleus. It is represented by symbol, II. The nucleus is, although, composed of neutrons and protons but it acts as if it is a single entity which has intrinsic angular momentum.

The nuclear spin depends on the mass number, if the mass number is odd then the nucleus has half-integer spin like the electron while if the nucleus has even mass number then its spin will be integer spin.

Nuclear Spin States

The nuclear spin states can take any number, fraction or integer. The number is dependent on the three points,

1. If both the neutrons and the protons in the nucleus are even in number then the nucleus has NO spin states.
2. If the sum of the neutrons and protons in the nucleus is odd then the nucleus has half integer spin (1/2. 3/2, 5/2, ...).
3. If both the neutrons and the protons in the nucleus are odd in number then the nucleus has an integer spin states (1, 2, 3, ...).

In other words the nucleus with odd number of protons or neutron or both should have the nuclear spin while if both are even then there is no nuclear spin.

Nuclear Spin Table

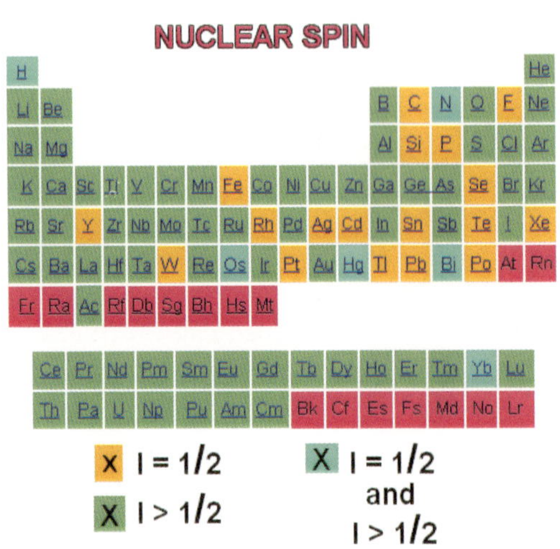

The nuclear spin of the various element of the periodic table is shown below. As shown the elements in the periodic tables are divided in three categories. The color code and their corresponding allowed nuclear spin is shown in the figure. There are certain elements for which the value of IIis unknown.

Quantum Number

Quantum Numbers are a set of values which describes the state of any electron including the orientation and type of orbital where it is likely to be found, its spin and its distance from the nucleus.

Nuclear Spin Quantum Number

The nuclear spin quantum number *l*, determines the allowed spin states of the nucleus and it is represented by; Allowed Spin States = $2l+1$,

In the case of hydrogen, $l = \frac{1}{2}$ and hence the allowed spin states of the nucleus is 2. Similarly if the value of *l* of an element is known we can find the allowed states of the nucleus of that atom.

Nuclear Spin Angular Momentum

The nuclear angular momentum is the angular momentum which is possessed by some nuclei. The nuclear spin angular momentum has two quantum number associated with it, a spin quantum number II and magnetic moment quantum number mImI. The magnitude of the nuclear spin angular momentum is given by;

$$\text{Magnitude of Nuclear Spin Angular Momentum} = \hbar\sqrt{I(I+1)}$$

and the z – component of the nuclear spin angular momentum is given by;

$$z - \text{component of the nuclear spin angular momentum} = m_I\hbar$$

The value of m_I is given by $-I{-}I \leq m_I \leq +I{+}I$. From this equation,

$$\text{The number of spin of a nucleus is} = 2I+1;$$

As we know that the nucleus has protons and neutrons and since protons and neutrons has $I = \frac{1}{2}$, so the allowed states of the numbers of spin of nucleus are 2.

As we know that the nuclei is the assembly of the protons and neutron. The entire assembly of protons and neutrons (nucleons) has a resultant angular momentum due to their angular momentum which usually is denoted by II. If the total angular momentum of a neutron is $j_n = \ell + s$ and for a proton is $j_p = \ell + s$ (where s for protons and neutrons happens to be $\frac{1}{2}$ same as that of electron) then the nuclear angular momentum quantum numbers II are given by:

$$I = |jn - jp|, |jn - jp| + 1, |jn - jp| + 2, \ldots\ldots, |jn - jp| - 2, |jn - jp| - 1, |jn - jp|$$

The nuclear spin angular momentum for the hydrogen is:

$_1H^1 \; I = \left(\frac{1}{2}\right)^+$

$_1H^2 \; I = 1^+$

$_1H^3 \; I = \left(\frac{1}{2}\right)^+$

Nuclear Spin Calculations

There are several methods for finding the nuclear spin of the given atom. The Stern-Gerlach experiment may be used to calculate the spin of the nucleus. This experiment is used to find the electron and atomic spin of the elements. The experiment was named after *Otto Stern* and *Walther Gerlach* and it is conducted by them in 1922. This was the first attempt to show that the elementary particles possess the quantum properties.

The Stern-Gerlach experimental apparatus consists essentially of a magnet which produces a non uniform magnetic field. A beam of atoms enters from the one side of the magnet in a direction perpendicular to the magnetic field. As a consequence of the interaction of nuclear and electronic spin with the applied magnetic field, the atoms undergo the deflection, as they go through the field. As they come out of the other side of the magnet, the atoms are detected by counters, possibly acting also as filters.

The particles (or atoms) spin is calculated by checking whether the particle (or atom) is deflected in the non uniform magnetic field. If the magnetic moment of the particle (or atom) is aligned with that of magnetic field than there is no deflection else it will deflect depending upon the net magnetic field on the particle (or atom). The deflection may be upwards or downwards.

This experiment showed that the electron (or atom) has some total angular momentum while earlier it was understood that the electron (or atom) has zero total angular momentum. If that was the case then the particles should not deflect when they are subjected to the magnetic field.

Nuclear Spin Resonance

The Nuclear Spin Resonance is study of the magnetic nuclei under the influence of the external magnetic field. The NSR is the phenomena used in many scientific instruments and experiment, in which a magnetic nucleus absorbs and emits the electromagnetic radiations under the influence of the external magnetic field.

In the normal condition the nuclear spin is randomly oriented but under the influence of the external magnetic field the nuclear spin either aligns in the same direction with the applied magnetic field or aligns in the opposite direction with the applied magnetic field. The nuclear spin aligned with the magnetic field has lower energy than the nuclear spin aligned opposite to the magnetic field.

Nuclear Stability

The nucleus is made of neutrons and protons. What causes them to stick together? Why do the protons not repel each other? Surprisingly, the mass of the nucleus is less than the sum of the masses of the individual protons and neutrons which makeup the nucleus. The lost mass (mass defect) has been changed into the energy necessary to bind the nucleus together. Suppose the mass defect is Δm. Then the energy binding the nucleus together ΔE is given by:

$$\Delta E = \Delta mc^2$$

where c is the speed of light. If is divided by the number of nucleons in the nucleus, then the binding energy per nucleon is obtained. The higher this value is the more stable is the nucleus.

The Nuclear Stability refers to the elements whose nucleus is stable, it means that the nuclear stability denotes the stability of the elements. Those atoms which have stable nucleus are known as nuclear stable atom.

Whenever we think of the radioactive material, we come across the concept of the nuclear stability and the isotopes. The elements are divided into two types on the basis of their nuclear stability:

1. Radioactive elements
2. Non-radioactive element

The non-radioactive element has stable nucleus and the radioactive element's nucleus is unstable. Hence, we can conclude that; Stable nucleus are non radioactive and Unstable nucleus are radioactive.

So, how can we be sure that the nucleus is stable by just examining the neutrons and protons of the element? Although there is no definitionsolution or answer to this question, there are several reasons for this question. We can understand by looking at the ratio of the neutrons and protons in nucleus of the several elements and there stability.

Protons	Neutrons	No. of stable nuclides	Stability
Odd	Even	50	least stable
Even	Odd	57	
Even	Even	168	most stable

From the above table we can conclude that:

1. Nuclides with odd number of protons and neutrons are least stable and more radioactive.
2. Nuclides with even number of protons and neutrons are most stable and hence least radioactive.

From the above discussion it is clear that if the number of both the protons to neutrons is even, then the nuclides tends to be more stable.

Nucleus Stability

If an atom of an element is non-radioactive it means that its nucleus is stable and does not disintegrate. Nucleus stability is makes certain isotopes radioactive. An isotope is said to be unstable, if the ratio of its protons to neutrons isn't within what is called the "Band of Stability." Elements with atomic numbers greater than 70 are never stable. Unstable isotopes generally undergo nuclear transmutation, alpha decay or beta decay.

Isotope Stability

Isotopes are the variants of a particular element. The elements may have 'n' number of isotopes. The isotopes of an element have same number of protons but differ in the number of neutrons.

The isotopes are said to be divided into two types:

1. Radioactive: The radioactive isotopes undergo radioactive decay and get disintegrated.
2. Stable: The isotopes stability defines by the stable nucleus of the element (isotope). The stable isotopes never observed to have undergone the process of nuclear disintegration.

Stable Isotope

The isotope with the stable nuclei and which have never undergone radioactive decay are said to be stable isotope. For e.g. consider the isotopes of carbon ^{14}C, ^{13}C, ^{12}C. From these isotopes of carbon ^{14}C is the radioactive isotope of carbon while the rest of the two are stable isotopes of carbon.

Nuclear Stability Graph

The nucleus holds the neutrons and protons by the forces which counter balances each other. These forces are proton to neutron binding force and the proton-proton repelling force. These forces counterbalance each other and hence keep the nucleus of the atom stable. In the diagram shownbelow, the red line depicts the stable nuclei and green line depicts the nuclei with the equal number of neutron and protons.

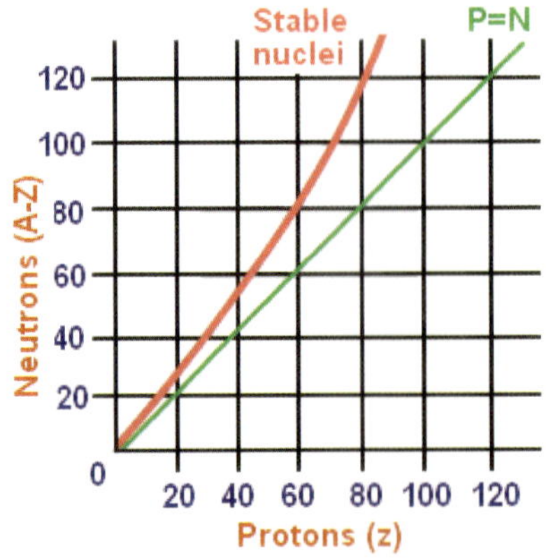

As can be seen from the diagram the stable nuclei could be found in atoms with atomic number (number of protons) no greater than 83 or the number of neutrons not greater than 126. As depicted in the graph above the more the protons are there in the nucleus of an atom the more number of neutrons are required to make the nucleus stable.

By further analyzing the graph it is understood that the stability of the nucleus is favoured the by the even number of electrons and neutrons in the nucleus. There are around 168 numbers of stable nuclei which have even number of neutrons and protons and only 4 stable nuclei which have odd number of both protons and neutrons.

The unstable nuclei decay giving away emission of elementary particle and/or energy. We will now describe different types of nuclear decay. Whenever nuclei disintegrate the atomic mass and atomic number remains same before and after the decay.

Alpha Decay

It involves the emission of alpha particle(s) and happens with the nuclei of atomic number less than 83. In this type of decay the unstable element (element with atomic number > 83) are shifted towards the more stable (red) belt of the above graph.

Beta Decay

It involves the conversion of the neutron into the proton and happens for the elements with high neutron to proton ratio. The elements shifts to the right on the stability graph.

Gamma Decay

It occurs with the other types of the decays. Generally the gamma decay doesn't affect the atomic mass & atomic number and hence the gamma particle is not shown in the gamma decay equation.

Positron Decay

It is the reverse phenomenon of the gamma decay. In this decay the proton of the nuclei turned into the neutron and happens with the elements of low neutron to proton ratio. The element after the decay is shifted up and towards the stability (red) belt.

Electron Capture

It is the process wherein the proton rich element absorbs the inner electron so that a proton is turned to the neutron. In this process a neutrino is emitted. It is the primary decay mode of the isotopic elements. It is the reverse phenomenon of the beta decay and hence sometimes named as inverse beta decay.

These all phenomenon are shown below by another diagram,

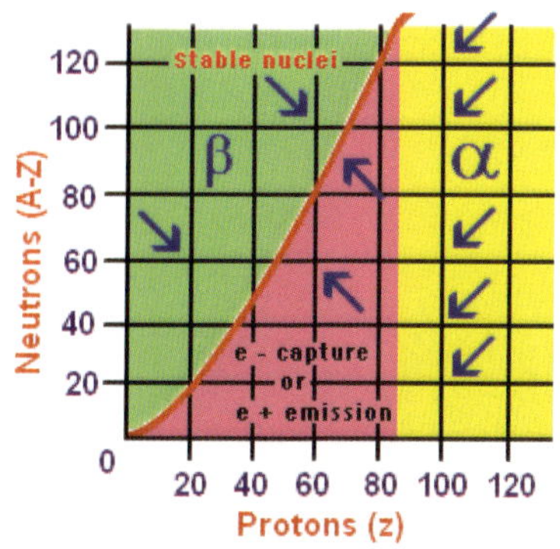

From this graph it is clear that the beta and alpha decay occur for different types of the elements. The beta decay occurs for the stable nuclei and alpha decay occurs for the unstable nuclei (atomic number > 83). Although most of the radioactive elements follows the above discussion, but there are few exceptions to it. For example, $_{92}U^{238}$ will go through the 8 alpha and 6 beta emissions before it becomes $_{82}Pb^{206}$.

Nuclear Reactors

A Nuclear Reactor is a device for obtaining and using the energy from a controlled nuclear chain reaction. Controlled means the rate or speed of the nuclear fissions can be changed by the operator. Most reactors are constructed in a similar manner. The six main parts of a nuclear reactor areNuclear Fuel, moderator, control rods, coolant, heat exchanger, and safety shields.

A Nuclear Reactor is used to extract the nuclear power from the radioactive material. The nuclear reactor initiates and controls the nuclear chain reaction. The present day nuclear reactors are based on the nuclear fission reaction.The Nuclear Reactor harness the thermal energy released by the nuclear fission reaction and this released energy is absorbed by the water or coolant and then it is converted to gas with high pressure. This high pressure gas is then passed to the gas turbine which generates the electricity from this high pressure gas.

Types

The following are the types of nuclear reactors.

Pressurized Water Reactor (PWR)

PWR is the most commonly used nuclear reactor. It uses water as a coolant. It has two loops of coolant water. The first or primary loop of water absorbs heat from the reactor. The water in the primary loop is kept at high pressure so that it doesn't boil.

The water in primary loop goes through the exchanger and the heat is transferred to the secondary loop coolant. The secondary loop coolant then spins the turbine.

Boiling Water Reactor (BWR)

BWR is very similar to that of the PWR and the only difference between the two is that in BWR the reactor has only one coolant loop. The water absorbs the heat from the reactor and the boiled water then spins the turbine.

Sodium Cooled Fast Reactor (SCF)

The SCF reactor uses sodium as a coolant. The water as a coolant acts as a neutron moderator which slows the fast neutrons and hence loss of energy. The liquid metal coolants provide thermal inertia against overheating.

Canada Deuterium-Uranium Reactors (CANDU)

The CANDU reactors are of Canadian design. It uses heavy water as a coolant. The heavy water (deuterium) absorbs lesser neutrons than the normal water atom. This reactor uses natural uranium instead of the enriched uranium required in most of the reactors.

High Temperature Gas Cooled Reactor (HTGR)

It is also known as very high temperature reactor (VHTR). As from the name it is evident that this reactor is gas cooled. These reactors are based on either prismatic block reactors or pebble bed reactors.

Liquid Fluoride Thorium Reactor

This is highly sophisticated form of the nuclear reactor. In this type of reactor the fuel itself is in the molten form. This helps in achieving high thermal efficiency while staying at low pressure vapor.

Working of a Nuclear Reactor

It works in the following steps:

1. Due to the fission reaction the heat produced is transferred to the water in the primary coolant tube.
2. The water in primary coolant tube exchange the heat with the coolant in the secondary coolant chamber (steam generator).
3. The steam generated from the above process is passed to the condenser chamber where it spins the turbine.
4. The turbine in turn connected to the generator which produces the electricity. This electricity is supplied to millions of home.

The Nuclear fission reactor works by bombarding the fuel (like uranium) by the neutron of very high speed. The reaction of the neutron and the uranium atom results in the disintegration of uranium atom into two atoms and a neutron. The extra neutron emitted due to this reaction is used for the next cycle of the fission reaction and this process continues as a chain reaction.

Due to this chain reaction excessive heat is produced and it is absorbed by the coolant of the reactor and is converted into the electricity by the turbine.

Components

The components and their functions are listed below:

1. Core: The Core contains the Nuclear Fuel and it produces most of the heat. The enriched or natural radioactive element (like, uranium) is as a fuel in most reactors.
2. Coolant: The coolant used mostly is water or deuterium. The coolant absorbs the heat produced by the core and uses this heat to spin the turbine to produce the electricity.

3. Turbine: It transfers the heat energy from the coolant to the electricity.
4. Containment: This is the steel reinforced concrete structure to isolate the reactor from the external environment.
5. Cooling towers: The role of cooling tower is to emit excessive heat in the form of water vapor. This is the part of the heat which cannot be converted into the electricity.

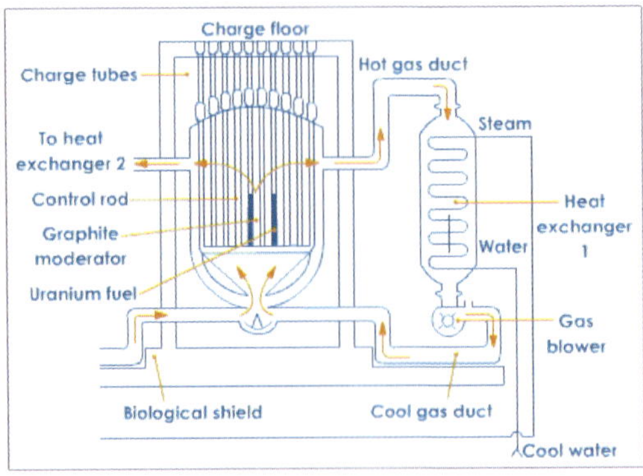

Small Nuclear Reactors

The Nuclear reactor which power at least 15% of world are of huge size and different types and based on different type of Nuclear Fuel and coolant technologies. The nuclear reactors are the future source of the world energy needs. The main problem with the nuclear power is the size of the reactor required to produce the electricity. These days more and more research is going on the development of a nuclear reactor which is not only smaller than the conventional nuclear reactor but also more fuel efficient.

Natural Nuclear Reactor

The Naturally occurring nuclear reactor does not exist in modern world. The natural nuclear reactor occurred on the earth about two million years ago. The natural deposit of the uranium acts as fission fuel and the underground water source acts as Neutron Moderator. The ancient natural nuclear reactor was found mostly in Africa.

The ancient nuclear reactor is self sustainable since there were huge deposits of the uranium available in earth. In modern days the uranium is not in huge quantities to self sustain the nuclear fission reaction naturally.

Nuclear Fission

Fission generally refers to splitting. In case of nuclear or atomic fission, it is division or splitting of atom or nucleus into sub atoms or sub nuclei. The generated atoms or nuclei are known as daughter nuclei or daughter atoms. And the original matter is called the parent. Fission is always followed by release of large amount of energy. Nuclear or atomic fission follow conservation of energy. Fission occurs naturally but very rarely.

Fission can be self repeated again and again so that it forms a chain which can be controlled or not, on the basis of which it is called controlled chain reaction and uncontrolled chain reaction. Chain reactions releases immense amount of energy. The energy released by fission is used for various purposes, to produce heat which can be used in different fields.

Fission can occur in two types of materials, fissile and fissionable. Fission process can be used in atomic or nuclear bombs, for generating heat which can be used as power supply, to activate some nuclear reactor or radioactive source, to produce nuclear fuels etc.

Fission is a process in which a nucleus or heavy nucleus or atom splits into lighter nuclei or atoms. In common terms, fission is a process of splitting something like an atom or cells into two or more sub atoms or cells. Fission process is accompanied by the release of large amount of energy.

Fission is mostly binary i.e. only two daughter nuclei are formed. In some cases fission occurs with the formation of three daughter nuclei or daughter elementary particles and this process is known as ternary fission. Fission occurs naturally as well as artificially. Fission can be seen in radioactive decay or in nuclear reactions.

The discovery of nuclear fission began to start with the discovery of neutron in the year 1932. Neutron was first discovered by *James Chadwick*. After some time *Enrico Fermi* started investigating the nuclear reactions of different atoms by bombardment with neutron.

After two years his coworkers observed that a minimum of different types of species were produced by bombarding uranium with a slow neutron. But the results from these observations were confusing and were clearly described and proved later in 1939 by *Otto Hahn* and *Fritz Strassmann* that the elements were actually barium, lanthanum and some other elements. The word fission for this process was first used by *Lise Meitner* and *Otto Frisch* in 1939.

Fission Reaction

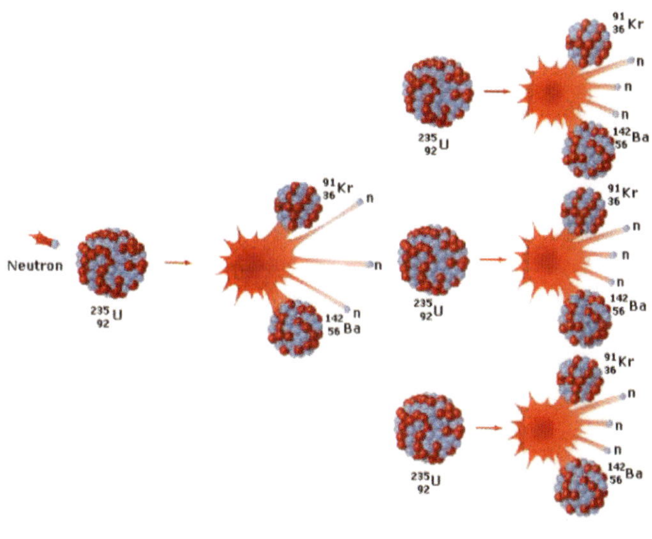

Nuclear Fission is the process of splitting a heavy nucleus to form two new elements with smaller nuclei. One of the most important nuclear reactions is that of uranium 235 which is used in nuclear reactors and atomic weapons. The fission of uranium 235 is initiated when a neutron of enough velocity collides with the nucleus of uranium 235. This collision produces uranium 236 which immediately splits into two lighter elements i.e. krypton 91 and barium 142. This event also releases three neutrons and energy. The neutrons produced are capable of producing more nuclear fission events if they encounter a uranium 235 nucleus. If a sufficient amount of mass of uranium 235 is present then the reaction becomes self sustaining resulting in the production of massive amount of energy, heat, light, ionizing radiation and atomic explosion. Nuclear Fission Equation of Uranium 235 is given below,

$$_{92}U^{235} + {}_{0}n^{1} \rightarrow 3\,{}_{0}n^{1} + {}_{36}Kr^{92} + {}_{56}Ba^{141} + \text{ENERGY}$$

Nuclear Fission Equation is basically a chemical equation which shows the production of new elements or nuclei from the original element or nucleus. For example:

$$_{94}Pu^{239} \rightarrow {}_{2}He^{4} + {}_{92}U^{235} \text{ (considered nuclear decay)}$$

Here, Plutonium is undergoing transformation into helium and uranium. These equations can be solved by balancing atomic number and mass number on both sides of the equation.

Fission Process

A Fission Process is possible without bombardment of neutron as we see in radioactive decay. This type of process occurs in heavy elements only. Nuclear fission occurs artificially by bombardment of mostly a neutron onto a nucleus. When a nucleus captures neutron it produces nuclear excitation energy. This energy is produced due to the nuclear force of attraction between neutron and nucleus. As a result of this nucleus gets deformed into a double-lobed drop, to a point that the distance between nuclear fragments increases such that the nuclear force cannot hold the charged nucleons.

As a result the fragments get separated and are moved away by the mutual charge of repulsion. This whole process occurs for fissile isotopes such as uranium 235 which undergo fission by slow neutrons. This process also occurs for fissionable material but they require more energy hence fast moving or high energy neutrons are required for their fission like uranium 238.

Atomic Fission

Atomic Fission is the process of splitting an atom into two or more atoms. If it splits into two then it is called binary atomic fission and if in three then called as ternary atomic fission. For example, uranium atom splitting into barium and krypton atoms is a binary atomic fission. Most of the atomic fission are binary. A very few are ternary fission. These Fission also involve release of energy. Due to this fission, some mass is lost conserving energy. A repeatedly occurring atomic fission or uncontrolled atomic fission results into explosion and is used as atomic bombs.

Cold Fission

As now we have seen that fission process involves release of neutrons but in cold fission no neutron or gamma rays are emitted. This is due to the low excitation energy of products. To study cold fission high flux nuclear reactors are used. Important fact of cold fission is that the mass of fission products is equal to the initial mass of the element.

Fission Chain Reaction

A Fission Chain Reaction is a propagated or continuous fission reaction in which the daughter nuclei that are produced by parent nucleus further undergoes or involve in another fission reaction.

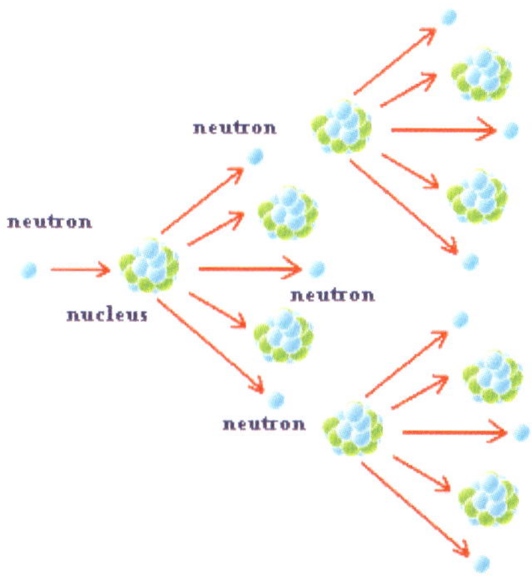

This can be explained with the example of uranium 235. Uranium 235 is bombarded with one neutron initially which results in fission of its nucleus and formation of krypton, barium and 3 neutrons. These three neutrons further bombards on other U-235 atoms if they are available in the sample of uranium 235. This reaction at second step further can cause nuclear fission and continues till the last atom of uranium is present. This phenomenon of self propagation of nuclear fission reaction is called Fission Chain Reaction or simply chain reaction.

This chain reaction was possible due to the release of more number of neutrons than were initially in the reaction. In case of uranium 238, fission occurs by bombardment of one neutron and emission of only one neutron. In this case chain reaction is not possible. Isotopes can undergo chain reaction only if they release more number of neutrons in fission. These isotopes are called fissionable. For example: uranium-235 and plutonium-239.

To support self sustaining chain reaction, certain minimum amount of fissionable matter is required. If the amount is less then there is possibility that the neutrons produced get emitted out of the sample without collision with other atom of the sample. If it happens than no more neutrons will be released and hence no fission occurs further and reaction stops. This minimum amount of material required for self sustaining chain reaction is called the critical mass.

A chain reaction can be of two types: controlled and uncontrolled. A controlled chain reaction is reaction that has some restrictions. A controlled chain reaction occurs for only some time and is only for the purpose of splitting a nucleus or for generation of certain amount of energy without resulting in an explosion. If the reaction gets out of control and continues very fast with the release of very immense amount of energy then it is called an uncontrolled reaction and this large amount of energy causes atomic explosion. All chain reactions need to be slowed down and controlled. This can be done by various methods.

Fission Energy

As we have known till now that fission occurs with release of large amount of energy which is known as fission energy. But the question arises that where does it come from. As we do an accurate measurement of the mass of the elements produced after fission and of the elements before fission we see that there is a difference in mass of the two. The mass after fission is less than it was initially. We can say that some mass disappeared during the fission. The loss of mass is known as mass defect. This lost mass is converted into energy. This is the energy released in fission.

We can measure the amount of energy released due to a fission process with mass energy equation given by Einstein. According to this equation,

$$E = \Delta mc^2$$

Here, E is the amount of energy released, Δm is the mass difference or lost mass or mass defect, and cc is the speed of light, which is 3×1083×108m/s. This results in the amount of energy released which was produced by the mass loss. We can say that some portion of mass is converted into energy.

Fission Products

Fission Products are those nuclides that are produced by a fission or nuclear reaction. Fission products are always more than the reactants as after fission a nucleus is split into two or more nuclides. The fission products are usually unstable in nature and due to this they go through beta decay and hence emit some more energy.

The summation of the atomic weights of the daughter atoms or fission products produced by fission is less than the atom that has undergone fission. This is due to the mass loss in the neutrons that are produced in the fission process. Also the parent nucleus always has more number of neutrons than the daughter nuclei.

Fissionable Material

A Fissionable material is that whose nuclides can undergo fission only when they are bombarded with a neutron. Fissionable material can undergo fission only by a high energy neutron. For example uranium - 238 is a fissionable material as it can undergo fission only by energy neutron. The binding energy of uranium 238 after bombardment with neutron is less than the critical energy. So a high energy neutron is required for fission. Examples of fissionable material: uranium – 238, thorium – 232.

Nuclear Fission Diagram

Nuclear Fission Diagram is simply a diagram showing the atom that undergoes fission and the fission products produced. The diagram shows the target nucleus, incident neutron, fission product and the neutrons produced.

Examples

Some other Examples of Nuclear Fission Equations are as follow.

$$_{92}U^{235} + {}_{0}n^{1} \rightarrow {}_{55}Cs^{140} + {}_{37}Rb^{93} + 3\,{}_{0}n^{1} + \text{Energy}$$

$${}_{92}U^{235} + {}_{0}n^{1} \rightarrow {}_{56}Ba^{141} + {}_{36}Kr^{92} + 3\,{}_{0}n^{1} + \text{Energy}$$

$${}_{92}U^{235} + {}_{0}n^{1} \rightarrow {}_{56}Ba^{137} + {}_{36}Kr^{97} + 3\,{}_{0}n^{1} + \text{Energy}$$

$${}_{92}U^{235} + {}_{0}n^{1} \rightarrow {}_{92}U^{236} \rightarrow {}_{54}Xe^{140} + {}_{38}Sr^{94} + 2\,{}_{0}n^{1} + \text{Energy}$$

Nuclear Fusion

Nuclear fusion, process by which nuclear reactions between light elements form heavier elements (up to iron). In cases where the interacting nuclei belong to elements with low atomic numbers (e.g., hydrogen [atomic number 1] or its isotopes deuterium and tritium), substantial amounts of energy are released. The vast energy potential of nuclear fusion was first exploited in thermonuclear weapons, or hydrogen bombs, which were developed in the decade immediately following World War II. Meanwhile, the potential peaceful applications of nuclear fusion, especially in view of the essentially limitless supply of fusion fuel on Earth, have encouraged an immense effort to harness this process for the production of power.

Fusion Reaction

Fusion reactions constitute the fundamental energy source of stars, including the Sun. The evolution of stars can be viewed as a passage through various stages as thermonuclear reactions and nucleosynthesis cause compositional changes over long time spans. Hydrogen (H) "burning" initiates the fusion energy source of stars and leads to the formation of helium (He). Generation of fusion energy for practical use also relies on fusion reactions between the lightest elements that burn to form helium. In fact, the heavy isotopes of hydrogen—deuterium (D) and tritium (T)—react more efficiently with each other, and, when they do undergo fusion, they yield more energy per reaction than do two hydrogen nuclei. (The hydrogen nucleus consists of a single proton. The deuterium nucleus has one proton and one neutron, while tritium has one proton and two neutrons.)

Fusion reactions between light elements, like fission reactions that split heavy elements, release energy because of a key feature of nuclear matter called the binding energy, which can be released through fusion or fission. The binding energy of the nucleus is a measure of the efficiency with which its constituent nucleons are bound together. Take, for example, an element with Z protons and N neutrons in its nucleus. The element's atomic weight A is $Z + N$, and its atomic number is Z. The binding energy B is the energy associated with the mass difference between the Z protons and N neutrons considered separately and the nucleons bound together ($Z + N$) in a nucleus of mass M. The formula is

$$B = (Zmp + Nmn - M)c^2,$$

where mp and mn are the proton and neutron masses and c is the speed of light. It has been determined experimentally that the binding energy per nucleon is a maximum of about $1.4\ 10^{-12}$ jouleat an atomic mass number of approximately 60—that is, approximately the atomic mass number of iron. Accordingly, the fusion of elements lighter than iron or the splitting of heavier ones generally leads to a net release of energy.

Two Types of Fusion Reactions

Fusion reactions are of two basic types: (1) those that preserve the number of protons and neutrons and (2) those that involve a conversion between protons and neutrons. Reactions of the first type are most important for practical fusion energy production, whereas those of the second type are crucial to the initiation of star burning. An arbitrary element is indicated by the notation *AZX*, where *Z* is the charge of the nucleus and *A* is the atomic weight. An important fusion reaction for practical energy generation is that between deuterium and tritium (the D-T fusion reaction). It produces helium (He) and a neutron (*n*) and is written,

$$\mathrm{D} + \mathrm{T} \rightarrow \mathrm{He} + n.$$

To the left of the arrow (before the reaction) there are two protons and three neutrons. The same is true on the right.

The other reaction, that which initiates star burning, involves the fusion of two hydrogen nuclei to form deuterium (the H-H fusion reaction):

$$\mathrm{H} + \mathrm{H} \rightarrow \mathrm{D} + \beta^{+} + \nu,$$

where β^{+} represents a positron and ν stands for a neutrino. Before the reaction there are two hydrogen nuclei (that is, two protons). Afterward there are one proton and one neutron (bound together as the nucleus of deuterium) plus a positron and a neutrino (produced as a consequence of the conversion of one proton to a neutron).

Both of these fusion reactions are exoergic and so yield energy. The German-born physicist Hans Bethe proposed in the 1930s that the H-H fusion reaction could occur with a net release of energy and provide, along with subsequent reactions, the fundamental energy source sustaining the stars. However, practical energy generation requires the D-T reaction for two reasons: first, the rate of reactions between deuterium and tritium is much higher than that between protons; second, the net energy release from the D-T reaction is 40 times greater than that from the H-H reaction.

Energy Released in Fusion Reactions

Energy is released in a nuclear reaction if the total mass of the resultant particles is less than the mass of the initial reactants. To illustrate, suppose two nuclei, labeled *X* and *a*, react to form two other nuclei, *Y* and *b*, denoted,

$$X + a \rightarrow Y + b.$$

The particles *a* and *b* are often nucleons, either protons or neutrons, but in general can be any nuclei. Assuming that none of the particles is internally excited (i.e., each is in its ground state), the energy quantity called the *Q*-value for this reaction is defined as,

$$Q = (m_x + m_a - m_b - m_y)c^2,$$

where the *m*-letters refer to the mass of each particle and *c* is the speed of light. When the energy value *Q* is positive, the reaction is exoergic; when *Q* is negative, the reaction is endoergic (i.e., absorbs energy). When both the total proton number and the total neutron number are preserved before and after the reaction (as in D-T reactions), then the Q-value can be expressed in terms of the binding energy B of each particle as,

$$Q = By + Bb - Bx - Ba.$$

The D-T fusion reaction has a positive Q-value of 2.8×10^{-12} joule. The H-H fusion reaction is also exoergic, with a Q-value of 6.7×10^{-14} joule. To develop a sense for these figures, one might consider that one metric ton (1,000 kg, or almost 2,205 pounds) of deuterium would contain roughly 3×10^{32}atoms. If one ton of deuterium were to be consumed through the fusion reaction with tritium, the energy released would be 8.4×10^{20} joules. This can be compared with the energy content of one ton of coal—namely, 2.9×10^{10} joules. In other words, one ton of deuterium has the energy equivalent of approximately 29 billion tons of coal.

Rate and Yield of Fusion Reactions

The energy yield of a reaction between nuclei and the rate of such reactions are both important. These quantities have a profound influence in scientific areas such as nuclear astrophysics and the potential for nuclear production of electrical energy.

When a particle of one type passes through a collection of particles of the same or different type, there is a measurable chance that the particles will interact. The particles may interact in many ways, such as simply scattering, which means that they change direction and exchange energy, or they may undergo a nuclear fusion reaction. The measure of the likelihood that particles will interact is called the cross section, and the magnitude of the cross section depends on the type of interaction and the state and energy of the particles. The product of the cross section and the atomic density of the target particle is called the macroscopic cross section. The inverse of the macroscopic cross section is particularly noteworthy as it gives the mean distance an incident particle will travel before interacting with a target particle; this inverse measure is called the mean free path. Cross sections are measured by producing a beam of one particle at a given energy, allowing the beam to interact with a (usually thin) target made of the same or a different material, and measuring deflections or reaction products. In this way it is possible to determine the relative likelihood of one type of fusion reaction versus another, as well as the optimal conditions for a particular reaction.

The cross sections of fusion reactions can be measured experimentally or calculated theoretically, and they have been determined for many reactions over a wide range of particle energies. They are well known for practical fusion energy applications and are reasonably well known, though with gaps, for stellar evolution. Fusion reactions between nuclei, each with a positive charge of one or more, are the most important for both practical applications and the nucleosynthesis of the light elements in the burning stages of stars. Yet, it is well known that two positively charged nuclei repel each other electrostatically—i.e., they experience a repulsive force inversely proportional to the square of the distance separating them. This repulsion is called the Coulomb barrier. It is highly unlikely that two positive nuclei will approach each other closely enough to undergo a fusion reaction unless they have sufficient energy to overcome the Coulomb barrier. As a result, the cross section for fusion reactions between charged particles is very small unless the energy of the particles is high, at least 10^4 electron volts (1 eV $\cong 1.602 \times 10^{-19}$ joule) and often more than 10^5 or 10^6 eV. This explains why the centre of a star must be hot for the fuel to burn and why fuel for practical fusion energy systems must be heated to at least 50,000,000 kelvins (K; 90,000,000 °F). Only then will a reasonable fusion reaction rate and power output be achieved.

The phenomenon of the Coulomb barrier also explains a fundamental difference between energy generation by nuclear fusion and nuclear fission. While fission of heavy elements can be induced by either protons or neutrons, generation of fission energy for practical applications is dependent on neutrons to induce fission reactions in uranium or plutonium. Having no electric charge, the neutron is free to enter the nucleus even if its energy corresponds to room temperature. Fusion energy, relying as it does on the fusion reaction between light nuclei, occurs only when the particles are sufficiently energetic to overcome the Coulomb repulsive force. This requires the production and heating of the gaseous reactants to the high temperature state known as the plasma state.

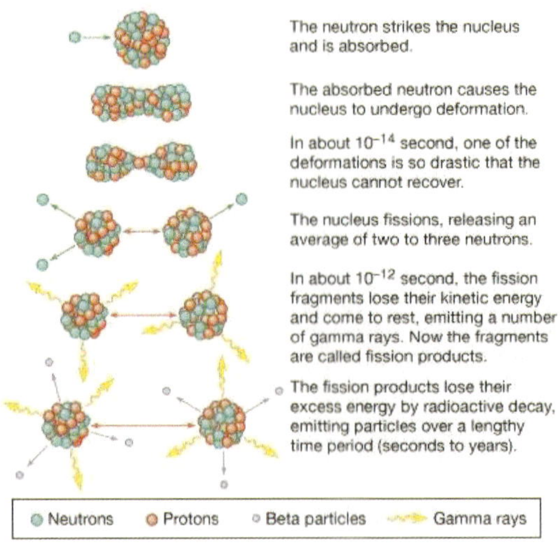

Plasma State

Typically, a plasma is a gas that has had some substantial portion of its constituent atoms or molecules ionized by the dissociation of one or more of their electrons. These free electrons enable plasmas to conduct electric charges, and a plasma is the only state of matter in which thermonuclear reactions can occur in a self-sustaining manner. Astrophysics and magnetic fusion research, among other fields, require extensive knowledge of how gases behave in the plasma state. The stars, the solar wind, and much of interstellar space are examples where the matter present is in the plasma state. Very high-temperature plasmas are fully ionized gases, which means that the ratio of neutral gas atoms to charged particles is small. For example, the ionization energy of hydrogen is 13.6 eV, while the average energy of a hydrogen ion in a plasma at 50,000,000 K is 6,462 eV. Thus, essentially all of the hydrogen in this plasma would be ionized.

A reaction-rate parameter more appropriate to the plasma state is obtained by accounting for the fact that the particles in a plasma, as in any gas, have a distribution of energies. That is to say, not all particles have the same energy. In simple plasmas this energy distribution is given by the Maxwell-Boltzmann distribution law, and the temperature of the gas or plasma is, within a proportionality constant, two-thirds of the average particle energy; i.e., the relationship between the average energy E and temperature T is $E = 3kT/2$, where k is the Boltzmann constant, 8.62×10^{-5} eV per kelvin. The intensity of nuclear fusion reactions in a plasma is derived by averaging the product of the particles' speed and their cross sections over a distribution of speeds corresponding to a Maxwell-Boltzmann distribution. The cross section for the reaction depends on the energy or speed of the particles. The averaging process yields a function for a given reaction that depends only on the temperature and can be denoted *f*(*T*). The rate of energy released (i.e., the power released) in a reaction between two species, *a* and *b*, is,

$$P_{ab} = n_a n_b f_{ab(T)} U_{ab},$$

where *na* and *nb* are the density of species *a* and *b* in the plasma, respectively, and *Uab* is the energy released each time *a* and *b* undergo a fusion reaction. The parameter *Pab* properly takes into account both the rate of a given reaction and the energy yield per reaction.

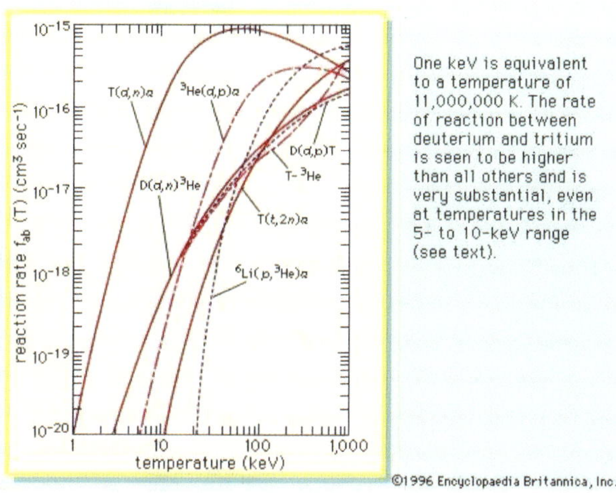

The reaction rate as a function of plasma temperature, expressed in kiloelectron volts (keV; 1 keV is equivalent to a temperature of 11,000,000 K). The rate of reaction between deuterium and tritium is seen to be higher than all others and is very substantial, even at temperatures in the 5-to-10-keV range

Fusion Reactions in Stars

Fusion reactions are the primary energy source of stars and the mechanism for the nucleosynthesisof the light elements. In the late 1930s Hans Bethe first recognized that the fusion of hydrogen nuclei to form deuterium is exoergic (i.e., there is a net release of energy) and, together with subsequent nuclear reactions, leads to the synthesis of helium. The formation of helium is the main source of energy emitted by normal stars, such as the Sun, where the burning-core plasma has a temperature of less than 15,000,000 K. However, because the gas from which a star is formed often contains some heavier elements, notably carbon (C) and nitrogen (N), it is important to include nuclear reactions between protons and these nuclei. The reaction chain between protons that ultimately leads to helium is the proton-proton cycle. When protons also induce the burning of carbon and nitrogen, the CN cycle must be considered; and, when oxygen (O) is included, still another alternative scheme, the CNO bi-cycle, must be accounted for.

The proton-proton nuclear fusion cycle in a star containing only hydrogen begins with the reaction,

$$H + H \rightarrow D + \beta^{+} + \nu; Q = 1.44 \text{ MeV},$$

where the *Q*-value assumes annihilation of the positron by an electron. The deuterium could react with other deuterium nuclei, but, because there is so much hydrogen, the D/H ratio is held to very low values, typically 10^{-18}. Thus, the next step is,

$$H + D \rightarrow {}^{3}He + \gamma;\ Q = 5.49\ \text{MeV},$$

where γ indicates that gamma rays carry off some of the energy yield. The burning of the helium-3 isotope then gives rise to ordinary helium and hydrogen via the last step in the chain:

$$^{3}He + {}^{3}He \rightarrow {}^{4}He + 2(H);\ Q = 12.86\ \text{MeV}.$$

At equilibrium, helium-3 burns predominantly by reactions with itself because its reaction rate with hydrogen is small, while burning with deuterium is negligible due to the very low deuterium concentration. Once helium-4 builds up, reactions with helium-3 can lead to the production of still-heavier elements, including beryllium-7, beryllium-8, lithium-7, and boron-8, if the temperature is greater than about 10,000,000 K.

The stages of stellar evolution are the result of compositional changes over very long periods. The size of a star, on the other hand, is determined by a balance between the pressure exerted by the hot plasma and the gravitational force of the star's mass. The energy of the burning core is transported toward the surface of the star, where it is radiated at an effective temperature. The effective temperature of the Sun's surface is about 6,000 K, and significant amounts of radiation in the visible and infrared wavelength ranges are emitted.

Fusion Reactions for Controlled Power Generation

Reactions between deuterium and tritium are the most important fusion reactions for controlled power generation because the cross sections for their occurrence are high, the practical plasma temperatures required for net energy release are moderate, and the energy yield of the reactions are high—17.58 MeV for the basic D-T fusion reaction.

It should be noted that any plasma containing deuterium automatically produces some tritium and helium-3 from reactions of deuterium with other deuterium ions. Other fusion reactions involving elements with an atomic number above 2 can be used, but only with much greater difficulty. This is because the Coulomb barrier increases with increasing charge of the nuclei, leading to the requirement that the plasma temperature exceed 1,000,000,000 K if a significant rate is to be achieved. Some of the more interesting reactions are:

$$H + {}^{11}B \rightarrow 3({}^{4}He);\ Q = 8.68\ \text{MeV};$$

$$H + {}^{6}Li \rightarrow {}^{3}He + {}^{4}He;\ Q = 4.023\ \text{MeV};$$

$$^{3}He + {}^{6}Li \rightarrow H + 2({}^{4}He);\ Q = 16.88\ \text{MeV; and}$$

$$^{3}He + {}^{6}Li \rightarrow D + {}^{7}Be;\ Q = 0.113\ \text{MeV}.$$

Reaction second converts lithium-6 to helium-3 and ordinary helium. Interestingly, if reaction second is followed by reaction third, then a proton will again be produced and be available to induce reaction second, thereby propagating the process. Unfortunately, it appears that reaction fourth is 10 times more likely to occur than reaction third.

Geiger–Nuttall Law

Geiger-Nuttall law is a semi-empirical law that expresses the lifetime (half-life) of the alpha emitter in terms of the energy of the released alpha particle. In other words, it states that short-lived isotopes emit more energetic alpha particles than long-lived ones. This rule was formulated by by Hans Geiger and John Mitchell Nuttall in 1911 prior to the development of the theoretical formulation. Geiger-Nuttall law can be mathematically expressed as:

$$\log\Gamma = \alpha + b\log(R\alpha)$$

where a and b are empirical constants that are found from logarithmic plots of experimental data. R_α represents the linear range of alpha particle, thus it is a direct measure of kinetic energy of the alpha particle. The width of the resonance (Γ) is generally related to the mean lifetime (τ) of the excited nucleus by the relation: $\Gamma = \hbar / \tau$

By 1928, George Gamow (and independently by Ronald Gurney and Edward Condon) had solved the theory of alpha decay via quantum tunneling. Using the tunneling mechanism, Gamow, Condon and Gurney calculated the penetrability of the tunneling α particle through the Coulomb barrier, finding the lifetimes of some α emitting nuclei. The main success of this model was the reproduction of the semi-empirical Geiger-Nuttall law that expresses the lifetimes of the α emitters in terms of the energies of the released α particles.

Internal Conversion Coefficient

The excess energy of radioactive nuclei in excited states is usually relieved through gamma emission. However, if the wave function of an orbital electron is such that it can exist close to or in the nucleus, the excess energy can be transferred directly to the orbital electron. Hence, as an alternative to gamma emission, the excited nucleus may return to a lower state or to the ground state by ejecting an orbital electron. This is known as internal conversion and results in the emission of a mono-energetic electron. Following the internal conversion, outer orbital electrons fill the deeper atomic energy levels, leading to characteristic X-ray emission or, alternatively, to the emission of Auger electrons.

The internal conversion process can be described by:

$$^{A}_{Z}P^* \rightarrow \left[^{A}_{Z}P\right]^* + e^-$$

The ratio of internal conversion to gamma emission photons is known as the internal conversion coefficient α where,

$$\alpha = \frac{N_e}{N\gamma} \text{SSS}$$

where N_e is the number of de-excitations via electron emission and $N\gamma$ the number of de-excitations via gamma emission. The kinetic energy E_e of the ejected electron is almost equal to the excitation energy E^* less the binding energy of the orbital electron E_B, i.e.

$$E_e = E^* - E_B \text{SS}$$

The internal conversion coefficent alpha depends on the atomic number and the excitation energy , i.e.

$$\alpha\alpha\frac{Z^3}{E^*}$$

Internal conversion is more important in heavy nuclei in particular in the decay of low-lying excited states (small E^*). Gamma emission predominated in light nuclei.

Example

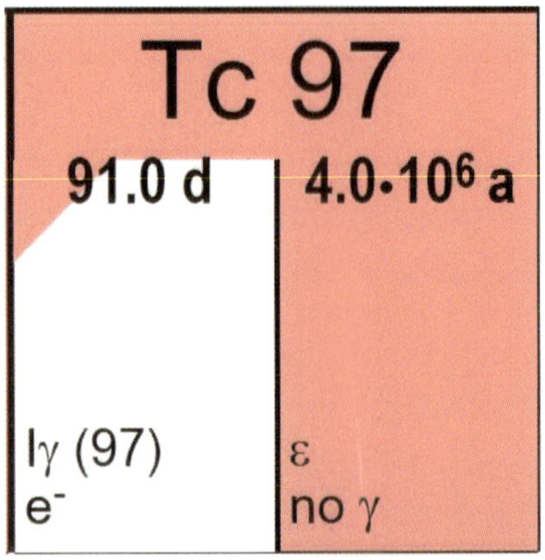

Tc97m in the Karlsruhe Nuclide Chart

Consider the decay of the isomeric state Tc97m (half-life 91 days). The box is mainly white indicating isomeric transition as the main decay mode and the small red triangle indicates electron capture to the daughter Mo97. The nuclide data shows Iγ (97) and e^-. The symbol e^- indicates that the converison coefficient is > 1 i.e. the state de-excites mainly by electron conversion rather than by gamma emission.

In addition, the fact that the energy of the isomeric transition is in backets i.e. (97) implies that gamma emission occurs in less that 1% of the de-excitations. Hence the electron conversion occurs is over 99% over the de-excitations. The actual conversion coefficient is not given in the chart, but can be obtained from the literature e.g nuclear Data sheets or from Nucleonica.

The conversion coefficient is α = 311 i.e 311 times more conversion electrons than gamma photons. These conversion electrons are seen in the beta spectrum. This information can be found in the references in Nucleonica. Together with the original research papers, information is given on branching ratios etc.

References

- "Nuclear ground-state properties and self-consistent calculations with the skyrme interaction". Nuclear Physics A. 238: 29–69. Bibcode:1975NuPhA.238...29B. doi:10.1016/0375-9474(75)90338-3

- Nuclear-magnetic-resonance, modern-physics: physics.tutorvista.com, Retrieved 31 March 2018
- Internal-conversion: nucleonica.com, Retrieved 11 July 2018
- Broglia, Ricardo A.; Zelevinsky, Vladimir (2013). Fifty Years of Nuclear BCS: Pairing in Finite Systems. World Scientific. ISBN 978-981-4412-48-3
- Nuclear-stability, modern-physics: physics.tutorvista.com, Retrieved 25 April 2018
- von Weizsäcker, C. F. (1935). "Zur Theorie der Kernmassen". Zeitschrift für Physik (in German). 96 (7–8): 431–458. Bibcode:1935ZPhy...96..431W. doi:10.1007/BF01337700
- Nuclear-fusion: britannica.com, Retrieved 26 May 2018

Permissions

All chapters in this book are published with permission under the Creative Commons Attribution Share Alike License or equivalent. Every chapter published in this book has been scrutinized by our experts. Their significance has been extensively debated. The topics covered herein carry significant information for a comprehensive understanding. They may even be implemented as practical applications or may be referred to as a beginning point for further studies.

We would like to thank the editorial team for lending their expertise to make the book truly unique. They have played a crucial role in the development of this book. Without their invaluable contributions this book wouldn't have been possible. They have made vital efforts to compile up to date information on the varied aspects of this subject to make this book a valuable addition to the collection of many professionals and students.

This book was conceptualized with the vision of imparting up-to-date and integrated information in this field. To ensure the same, a matchless editorial board was set up. Every individual on the board went through rigorous rounds of assessment to prove their worth. After which they invested a large part of their time researching and compiling the most relevant data for our readers.

The editorial board has been involved in producing this book since its inception. They have spent rigorous hours researching and exploring the diverse topics which have resulted in the successful publishing of this book. They have passed on their knowledge of decades through this book. To expedite this challenging task, the publisher supported the team at every step. A small team of assistant editors was also appointed to further simplify the editing procedure and attain best results for the readers.

Apart from the editorial board, the designing team has also invested a significant amount of their time in understanding the subject and creating the most relevant covers. They scrutinized every image to scout for the most suitable representation of the subject and create an appropriate cover for the book.

The publishing team has been an ardent support to the editorial, designing and production team. Their endless efforts to recruit the best for this project, has resulted in the accomplishment of this book. They are a veteran in the field of academics and their pool of knowledge is as vast as their experience in printing. Their expertise and guidance has proved useful at every step. Their uncompromising quality standards have made this book an exceptional effort. Their encouragement from time to time has been an inspiration for everyone.

The publisher and the editorial board hope that this book will prove to be a valuable piece of knowledge for students, practitioners and scholars across the globe.

Index